Environmental Impacts of Modern Agriculture

ISSUES IN ENVIRONMENTAL SCIENCE AND TECHNOLOGY

EDITORS:

R.E. Hester, University of York, UK
R.M. Harrison, University of Birmingham, UK

TITLES IN THE SERIES:

How to obtain future titles on publication

A subscription is available for this series. This will bring delivery of each new volume immediately on publication and also provide you with online access to each title via the Internet. For further information visit http://www.rsc.org/issues or write to the address below.

For further information please contact:
Sales and Customer Care, Royal Society of Chemistry, Thomas Graham House, Science Park, Milton Road, Cambridge, CB4 0WF, UK
Telephone: +44 (0)1223 432360, Fax: +44 (0)1223 426017, Email: sales@rsc.org

ISSUES IN ENVIRONMENTAL SCIENCE AND TECHNOLOGY

EDITORS: R.E. HESTER AND R.M. HARRISON

34
Environmental Impacts of Modern Agriculture

RSCPublishing

ISBN: 978-1-84973-385-4
ISSN: 1350-7583

A catalogue record for this book is available from the British Library

Published by The Royal Society of Chemistry,
Thomas Graham House, Science Park, Milton Road,
Cambridge CB4 0WF, UK

Registered Charity Number 207890

For further information see our web site at www.rsc.org

Printed and bound in Great Britain by CPI Group (UK) Ltd, Croydon, CR0 4YY, UK

Preface

Agriculture is something which the majority of us take for granted. Provided the usual items turn up on the supermarket shelves, there is little reason for most of the population to give the matter any thought. Yet, agriculture plays a critical role in feeding the world's population, and in many areas farmers are key custodians of the environment. Despite its predominantly rural setting, agriculture, and especially intensive agriculture, is a highly unnatural activity which substantially perturbs the environment, causing a wide range of both positive and negative consequences. In this volume many of those consequences are explored by expert authors.

The volume starts with a chapter by Joe Morris and Paul Burgess which considers the implications of agriculture for land use and management. With growing urbanisation, agriculture is expected to feed an ever-growing population from a diminishing area of available land. The chapter gives a broad view of the issues and challenges facing agriculture at the global scale and reviews the case of modern agriculture in the UK with particular reference to factors affecting land use. It also explores how agricultural technologies can help improve the productivity of farming systems while simultaneously delivering a range of other ecosystem benefits. Agricultural production is critically dependent upon good soil quality, and the second chapter by Ruben Sakrabani and his colleagues explores how agriculture impacts upon soil quality. Not only does agriculture affect the composition of soils, it can lead to compaction which is related to factors such as soil texture, packing density, moisture content and plasticity. The importance of soil biodiversity in maintaining the functionality of soils, and how this can be affected by soil contaminants is described.

While it is well known that agricultural production is a sink for carbon dioxide through photosynthesis, it is less well recognised that agriculture can also be a source of greenhouse gases. Soil respiration processes can release both carbon dioxide and methane, and nitrogenous fertilisers can be converted to nitrous oxide, a gas with a high global warming potential. In the third chapter, Joanna Cloy and a number of distinguished co-authors from across

Issues in Environmental Science and Technology, 34
Environmental Impacts of Modern Agriculture
Edited by R.E. Hester and R.M. Harrison
© The Royal Society of Chemistry 2012
Published by the Royal Society of Chemistry, www.rsc.org

the United Kingdom give a quantitative view on the effect of agriculture upon greenhouse gases and explain the opportunities which exist to mitigate emissions through modifications to agricultural management practices that involve more efficient use of inputs and promotion of carbon sequestration. The need for improved greenhouse gas budgets is also emphasised. Agriculture also leads to both chemical and microbiological pollution of surface and groundwaters, and in Chapter 4 David Kay and a number of his colleagues discuss the impacts of agriculture on water-borne pathogens. Livestock production is a major source of microbial pollution which is especially difficult to control as it is a highly diffuse source. This chapter outlines the problem and its significance as well as highlighting simple but effective mitigation measures.

Modern agriculture depends heavily upon the use of pesticides to maximise production and avoid potentially catastrophic crop losses due to insects, fungi and other organisms. The use of agricultural pesticides has the potential to damage sensitive ecosystems as well as exposing human populations to toxins. In Chapter 5, Daniel Osborn sets out the problems caused historically by the use of pesticides and the measures that are now taken to minimise harm to human health and the environment.

The final two chapters deal with broad but important issues which currently affect agriculture. In Chapter 6, Ian Crute examines – in an historical and contemporary context – some of the issues which arise in attempting to achieve a balance between producing more food and the adverse environmental impacts of agricultural systems. The chapter takes an holistic view of the benefits and drawbacks of modern agriculture and explores the concept of "sustainable intensification" as it might assist in addressing biodiversity, conservation and the need to mitigate global climate change by reducing net greenhouse gas emissions from agricultural systems. In the final chapter, Lucas Reijnders takes a balanced look at the positive and negative aspects of agricultural production of biofuels. This is currently a highly contentious topic with both governmental and commercial interests advocating greater use of renewable fuels, including biofuels, but large sectors of the scientific community pointing out the substantial adverse environmental impacts of intensive biofuel production.

Overall, the volume provides a selective but broadly based overview of current issues on a topic which continues to stimulate a great deal of active research despite having confronted society for very many years. We are grateful to the many distinguished authors who have contributed to this volume which we believe will be of immediate and lasting value, not only to practitioners in government, consultancy and industry, but also to environmentalists, policymakers and students taking courses in environmental science, agriculture, conservation and environmental management.

Ronald E. Hester
Roy M. Harrison

Contents

Issues in Environmental Science and Technology, 34
Environmental Impacts of Modern Agriculture
Edited by R.E. Hester and R.M. Harrison
© The Royal Society of Chemistry 2012
Published by the Royal Society of Chemistry, www.rsc.org

Editors

Ronald E. Hester, BSc, DSc (London), PhD (Cornell), FRSC, CChem

Ronald E. Hester is now Emeritus Professor of Chemistry in the University of York. He was for short periods a research fellow in Cambridge and an assistant professor at Cornell before being appointed to a lectureship in chemistry in York in 1965. He was a full professor in York from 1983 to 2001. His more than 300 publications are mainly in the area of vibrational spectroscopy, latterly focusing on time-resolved studies of photoreaction intermediates and on biomolecular systems in solution. He is active in environmental chemistry and is a founder member and former chairman of the Environment Group of the Royal Society of Chemistry and editor of 'Industry and the Environment in Perspective' (RSC, 1983) and 'Understanding Our Environment' (RSC, 1986). As a member of the Council of the UK Science and Engineering Research Council and several of its sub-committees, panels and boards, he has been heavily involved in national science policy and administration. He was, from 1991 to 1993, a member of the UK Department of the Environment Advisory Committee on Hazardous Substances and from 1995 to 2000 was a member of the Publications and Information Board of the Royal Society of Chemistry.

Roy M. Harrison, BSc, PhD, DSc (Birmingham), FRSC, CChem, FRMetS, Hon MFPH, Hon FFOM, Hon MCIEH

Roy M. Harrison is Queen Elizabeth II Birmingham Centenary Professor of Environmental Health in the University of Birmingham. He was previously Lecturer in Environmental Sciences at the University of Lancaster and Reader and Director of the Institute of Aerosol Science at the University of Essex. His more than 300 publications are mainly in the field of environmental chemistry, although his current work includes studies of human health impacts of atmospheric pollutants as well as research into the chemistry of pollution phenomena. He is a past Chairman of the Environment Group of the Royal Society of Chemistry for whom he has edited 'Pollution: Causes, Effects and Control' (RSC, 1983;

Fourth Edition, 2001) and 'Understanding our Environment: An Introduction to Environmental Chemistry and Pollution' (RSC, Third Edition, 1999). He has a close interest in scientific and policy aspects of air pollution, having been Chairman of the Department of Environment Quality of Urban Air Review Group and the DETR Atmospheric Particles Expert Group. He is currently a member of the DEFRA Air Quality Expert Group, the Department of Health Committee on the Medical Effects of Air Pollutants, and Committee on Toxicity.

List of Contributors

1. *Modern Agriculture and Implications for Land Use and Management*

Joe Morris (corresponding author), Resource Economics and Management, B42A, Department of Natural Resources, School of Applied Sciences, Cranfield University, Bedfordshire, MK43 0AL, UK, E-mail: J.Morris@cranfield.ac.uk

Paul J. Burgess, Resource Economics and Management, B42A, Department of Natural Resources, School of Applied Sciences, Cranfield University, Bedfordshire, MK43 0AL, UK, Email: P.Burgess@cranfield.ac.uk

2. *Impacts of Agriculture upon Soil Quality*

Ruben Sakrabani (corresponding author), Lecturer in Soil Chemistry, Building 40, Department of Environmental Science and Technology, School of Applied Sciences, Cranfield University, Cranfield, Bedfordshire, MK43 0AL, UK, Email: r.sakrabani@cranfield.ac.uk

Lynda Deeks, Building 40, Department of Environmental Science and Technology, School of Applied Science, Cranfield University, Cranfield, Bedfordshire, MK43 0AL, UK, Email: l.k.deeks@cranfield.ac.uk

Mark Kibblewhite, Building 40, Department of Environmental Science and Technology, School of Applied Science, Cranfield University, Cranfield, Bedfordshire, MK43 0AL, UK, Email: m.kibblewhite@cranfield.ac.uk

Karl Ritz, Building 40, Department of Environmental Science and Technology, School of Applied Science, Cranfield University, Cranfield, Bedfordshire, MK43 0AL, UK, Email: k.ritz@cranfield.ac.uk

3. *Impacts of Agriculture upon Greenhouse Gas Budgets*

Joanna M. Cloy (corresponding author), Scottish Agricultural College, West Mains Road, Kings Buildings, Edinburgh EH9 3JG, UK, Email: Joanna.Cloy@sac.ac.uk

Robert M. Rees, Scottish Agricultural College, West Mains Road, Kings Buildings, Edinburgh EH9 3JG, UK, Email: Bob.Rees@sac.ac.uk

Keith A. Smith, School of GeoSciences, University of Edinburgh, West Mains Road, Edinburgh, EH9 3JN, UK, Email: keith.smith@ed.ac.uk

Keith W.T. Goulding, Rothamsted Research (BBSRC), Harpenden, Hertfordshire, AL5 2JQ, UK, Email: keith.goulding@bbsrc.ac.uk

Peter Smith, School of Biological Sciences, University of Aberdeen, St Machar Drive, Aberdeen, AB24 3UU, UK, Email: pete.smith@abdn.ac.uk

A. Waterhouse, Scottish Agricultural College, West Mains Road, Kings Buildings, Edinburgh EH9 3JG, UK, Email: tony.waterhouse@sac.ac.uk

David Chadwick, Rothamsted Research (BBSRC), North Wyke Research, Okehampton, Devon, EX20 2SB, UK, Email: david.chadwick@bbsrc.ac.uk

4. *Impacts of Agriculture on Water-borne Pathogens*

David Kay (corresponding author), Institute of Geography & Earth Sciences, Centre for Research into Environment and Health Aberystwyth University, Aberystwyth, Ceredigion, SY23 3DB, UK, Email: dave@crehkay.demon.co.uk

John Crowther, Centre for Research into Environment and Health, University of Wales, Trinity St David, Lampeter, Ceredigion, SA48 7ED, UK

Cheryl Davies, Institute of Geography & Earth Sciences, Centre for Research into Environment and Health, Aberystwyth University, Aberystwyth, Ceredigion, SY23 3DB, UK

Tony Edwards, Centre for Research into Environment and Health Analytical Ltd, Hoyland House, 50 Back Lane, Horsforth, Leeds LS18 4RF, UK

Lorna Fewtrell, Institute of Geography & Earth Sciences, Centre for Research into Environment and Health, Aberystwyth University, Aberystwyth, Ceredigion, SY23 3DB, UK

Carol Francis, Centre for Research into Environment and Health Analytical Ltd, Hoyland House, 50 Back Lane, Horsforth, Leeds LS18 4RF, UK

Chris Kay, Institute of Geography & Earth Sciences, Centre for Research into Environment and Health, Aberystwyth University, Aberystwyth, Ceredigion, SY23 3DB, UK

Adrian McDonald, School of Earth and Environment, University of Leeds, Leeds S2 9JT, UK

Carl Stapleton, Institute of Geography & Earth Sciences, Centre for Research into Environment and Health, Aberystwyth University, Aberystwyth, Ceredigion, SY23 3DB, UK

John Watkins, Centre for Research into Environment and Health Analytical Ltd, Hoyland House, 50 Back Lane, Horsforth, Leeds LS18 4RF, UK

Mark Wyer, Institute of Geography & Earth Sciences, Centre for Research into Environment and Health, Aberystwyth University, Aberystwyth, Ceredigion, SY23 3DB, UK

5. *Pesticides in Modern Agriculture*

Dan Osborn, Natural Environment Research Council, Polaris House, North Star Avenue, Swindon, Wiltshire, SN2 1EU, UK, Email: dano@nerc.ac.uk

6. *Balancing the Environmental Consequences of Agriculture with the Need for Food Security*

Ian Crute, Agriculture and Horticulture Development Board, Stoneleigh Park, Kenilworth, Warwickshire, CV8 2TL, UK, Email: ian.crute@ahdb. org.uk

7. *Positive and Negative Impacts of Agricultural Production of Liquid Biofuels*

Lucas Reijnders, Institute for Biodiversity and Ecosystem Dynamics, University of Amsterdam, Science Park 904, P.O. Box 94248, 1090 GE Amsterdam, The Netherlands, Email: L.Reijnders@uva.nl

Modern Agriculture and Implications for Land Use and Management

JOE MORRIS* AND PAUL J. BURGESS

ABSTRACT

Agriculture is a land-based primary industry that directly depends on natural resources such as land, water, and a diversity of plants and animals. It is supported by the application of human knowledge, both traditional and scientific, and human effort, skills and endeavour. Following the global food and energy crisis in 2007–2008, and growing awareness of the challenges posed by climate change, interest in agriculture, after a period of neglect, has been reinstated as a key area of international and national policies. The performance of agriculture, however, must now be measured not only in terms of food, fibre and bio-energy production, but also a range of other social and environmental outcomes, positive and negative.

An array of technological, environmental, economic, social, and political factors, many of which are very context and spatially specific, shape the characteristics of agricultural systems, and the demand for, supply of and use of agricultural land. The emerging consensus is that agriculture must meet the needs of a growing and potentially more prosperous global population mainly from the existing stock of agricultural land. Otherwise the world's ecosystems could be irreparably damaged. Harnessing the potential of sustainable agricultural technologies to improve the productivity of farmed land while protecting the natural environment is a key element of this process, especially in regions where agricultural development is the main means of alleviating hunger and poverty.

Following a broad review of issues and challenges facing agriculture at the global scale, the chapter considers the case of modern agriculture in

---

*Corresponding author

Issues in Environmental Science and Technology, 34
Environmental Impacts of Modern Agriculture
Edited by R.E. Hester and R.M. Harrison
© The Royal Society of Chemistry 2012
Published by the Royal Society of Chemistry, www.rsc.org

the UK with particular reference to factors affecting land use. It also explores how agricultural technologies can help improve the productivity of farming systems while simultaneously delivering a range of other ecosystem benefits.

1   Introduction and Overview

Agriculture is a human activity carried out primarily to produce food, fibre and bio-energy by the deliberate and controlled use of (mainly terrestrial) plants and animals.[1] This process of crop production uses the process of photo-synthesis to capture solar energy and requires land and the associated soil and climate to provide water, nutrients, and anchorage. Animal production in turn is typically based on domesticated animals digesting plant material and pro-ducing high value products such as meat, milk and eggs. From a policy per-spective, the main aim of modern agriculture is to achieve a stable and affordable food supply for all, whilst reducing climate change and maintaining biodiversity and ecosystem services.[2] From a farmer's perspective, a principal aim is to secure a stable and rewarding livelihood.

Agricultural activity creates a land-based ecosystem that is focused on the commercial interaction of biotic resources (primarily domesticated crops and animals, but also soil micro-organisms), with abiotic resources such as the atmosphere, soil, water, and energy. It is supported by the application of human knowledge, whether traditional or scientific, and human effort, skills and endeavour. The performance of agriculture can be measured not only in terms of the economic value of agricultural outputs per unit of land ($ ha^{-1}) but also per unit of other inputs such as labour and finance. Increasingly there is awareness that performance also needs to consider a range of wider social and environmental outcomes which can be either positive or negative. Hence performance depends on a range of technological, environmental, economic, social, and political factors, which can be context and spatially specific.

This chapter reviews the characteristics of agricultural systems and the fac-tors operating at the global and regional scale that affect the demand for, supply of and use of agricultural land. The links between modern agriculture and land use are explored with respect to the case of the UK. As the demands on land increase, agricultural activities (at both regional and global scale) will need to combine the challenge of increased food production whilst protecting and enhancing the natural environment and this will also need to be done in the context of the effects of climate change. Harnessing the potential of sustainable agricultural technologies to augment the capacity of the existing stock of agricultural land is critical to this process.

2   Agricultural Systems

Agriculture can be viewed as a system: an organised assembly of components brought together for a purpose (Figure 1). Inputs such as land, genetic

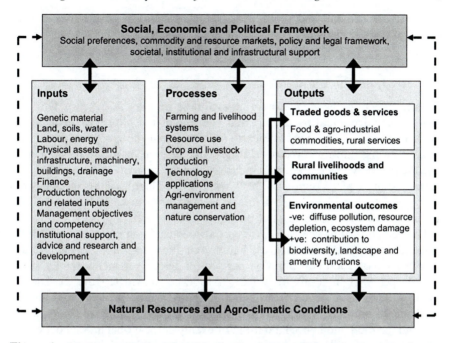

Figure 1 Modern agriculture functions in the context of natural resource and agro-climatic conditions and a socio-economic and political framework.

resources, labour, energy, water, capital, and human knowledge are combined in a range of farming systems, such as cropping or livestock systems, to produce a diverse array of outputs, some intended and some unintended. Climate, the soil environment, and topography often constrain the particular crops and animals that can be used in a particular area. However agricultural systems are also strongly influenced by political and institutional factors, such as agricultural policy and land tenure arrangements. Whereas, the dominant agricultural purpose is to produce food and biomass for human consumption, increasingly agriculture is charged with (i) contributing to a wider array of beneficial goods and services associated with the occupancy and use of the rural landscape and (ii) controlling the unintended negative consequences of modern farming methods. This requires an understanding of the synergies and tradeoffs between agriculture, the natural environment and human well-being. It also begs a wider question – 'what is land for?'.[3]

3   Global and Regional Issues

Two recent reviews of world agriculture, the International Assessment of Agricultural Knowledge, Science and Technology for Agricultural Development (IAASTD)[4] and the Foresight review on the Future of Food and Farming[5] examined the key drivers affecting global food and farming systems.

They include demand and supply side factors, systems of governance and the effects of climate change.

3.1 Demand Side Factors

Factors affecting the demand for agricultural outputs and hence land use include global population change and changes in the nature of demand for food, fibre and bio-energy. Forecasts for the global human population, predicated on assumptions about GDP growth, income distribution, education and the role of women, predict an increase from the current 7 billion people to a central estimate of 9.3 billion for 2050, with low and high variants ranging from 8.1 to 10.6 billion.[6] The current rate of population growth represents a 1.1% annual increase in the number of people requiring food.

The demand for agricultural products also depends on consumption habits, influenced by tastes, preferences, habits and lifestyles. With rising average global incomes, food consumption per person per year is increasing by about +0.2% per year in terms of weight of food consumed (Figure 2). Dietary changes, such as a switch from starch staples to fats, proteins and sugars, particularly amongst relatively low income groups, could radically affect demand for food and hence land use. For example, the global annual per capita consumption of milk is predicted to increase from 78 kg in 2000 to 115 kg in 2050. Moreover the annual per capita demand for meat is predicted to increase from 37 kg in 2000 to 52 kg in 2050.[7] Each kg increase in milk or meat requires an additional 5 to 8 kg of animal feed. Less than half of global grain production is consumed directly by humans, and this share could decrease as a result of global dietary change.[7]

While increasing food availability is a Millennium Development Goal for the 1 billion people that are currently undernourished, diet-related obesity and

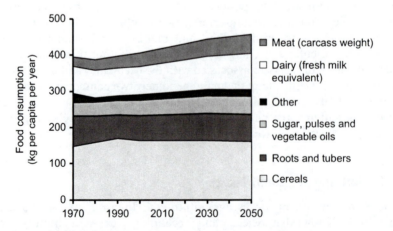

Figure 2   Predicted global commodity consumption per person by major food groups from 1970 to 2000 (actual data) and from 2010–2050 (predicted data).[7]

ill-health amongst 1 billion of the better off is also a concern.[8] Artificially high market specifications and avoidable food waste can also place further demand pressure on agriculture. For example, it is estimated that between a quarter and a third of all food for human consumption is probably lost in the world food system.[9,10] Nonetheless, the Food and Agriculture Organisation (FAO) estimates that food production needs to increase by 70% from 2010 to feed 9 billion people in 2050.[11]

Agriculture is also affected by the global energy demand that is predicted to increase by more than 50% in the next 25 years.[12] While agriculture can contribute bio-energy sources, modern agriculture is fossil energy intensive in its use of fertilisers and fuels and is therefore vulnerable to high energy prices.[13]

3.2 Supply Side Factors

On the supply side, agriculture's ability to meet demand depends on the availability and suitability of land (including associated hydrological and climatic factors), technologies, traditional and modern, that determine the productivity of farming systems, and the logistics, processing and marketing chains that link supply with end users.

Of the estimated 13 400 million ha of land on earth, one third, about 4600 million ha, are used for agriculture.[5] Of the total area, about 3000 million ha is considered suitable for crop production, of which two thirds are moderately suited to cereal production. At present, about half of the potential crop area (1400–1600 million ha) is cultivated for crops, the rest is mainly grassland. In the last 40 years (1967–2007), the total area of land occupied by agriculture has increased by only 8% but this has been mainly at the expense of forests, savannah and natural grasslands. Global yields increased by about 115% over the period, outpacing the 86% increase in world population, except for Africa where production per capita has only recently recovered to 1960 levels.[14] Average agricultural area declined during the period from 1.30 to 0.72 ha per person. Every year, an estimated 12 million hectares of agricultural land are lost to land degradation, adding to the billions of hectares that are already degraded.[15]

There are considerable regional variations in the absolute and relative changes in agricultural land use reflecting, in part, differences in governance, development and population growth (Figure 3). In Europe, for example most of the output has been achieved through increased yields on a declining agricultural area. Following rapid expansion of cropped areas in Asia in the 1960s and 1970s, much of it involving forest clearance, more recent increases in agricultural output have occurred on a declining area. In Africa and Latin America, the agricultural area is increasing, associated with reductions in forest and woodland areas. In Oceania, the large decline in the reported agricultural area (primarily of grazing land) was associated with an increase in designated conservation areas in Australia.[16]

Joe Morris and Paul J. Burgess

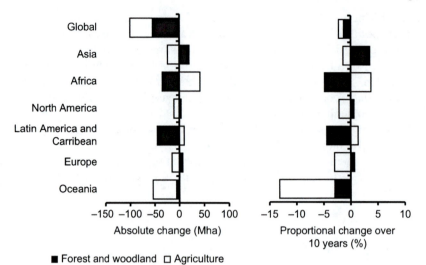

Figure 3 Absolute and relative changes in land occupied by agriculture and forestry/
woodland by major global regions 1999–2009 (derived from FAOSTAT,
2011).[17] (Proportional change is expressed as a proportion of the area of
agriculture or forest in 1999.)

Agriculture is dependent on water supply, whether naturally available 'green'
water from rainfall or 'blue' water artificially extracted from surface and
groundwater sources.[18] In drier regions, agriculture can account for up to 70%
of blue water withdrawals. Demand for irrigation water could double by 2050,
which, combined with declining available water due to climate change and
other demands on water resources, could constrain agricultural potential.[19] In
addition, agriculture will need to contend with changing levels and distribution
of rainfall and temperatures associated with climate change, including
increased incidence of extreme drought or flood events.[20] While some areas will
experience negative effects, others, particularly in temperate and boreal zones,
may derive benefit from warmer climates, especially if soils and water resources
are favourable. Agriculture also contributes to climate change and policies to
address these impacts, such as control of carbon emissions, also have major
implications for agriculture and land use.

Also on the supply side, agricultural commodity markets that induce global
land use are now much more internationally integrated, reflecting a reduction
in trade barriers, improvements in information technology and the actions of
transnational companies operating at the global scale. Emergent economies,
such as Brazil, Russia, India, and China now exert stronger influence on
agricultural markets, whether as importers or exporters. However, there
remains a relatively high degree of protection for the domestic markets of
developed countries, although much of this takes the form of indirect producer
subsidies rather than direct restrictions on trade. The trade restrictions imposed
following the 2008 spikes in global food prices, however, showed a limited

commitment to collaboration when national interests are threatened. Protectionist and market distorting policies mean that supplies and prices in residual markets become more volatile with consequences for the poorest nations.

Global food and energy shortages have persuaded many governments to re-examine their approach to food security and the importance of maintaining a strategic capacity in production at a national or regional scale. Estimates of the effect of biofuels on the commodity price spike in 2007 range from 30%[21] to 75%.[22] In future, the additional land required for biofuel production could displace food production, leading to food shortages and higher food prices.[23,24]

4 Agricultural Land and the Role of Science and Technology

Although the application of science and technology has led to major improvements in agricultural performance, poverty, hunger and environmental degradation are still widespread.

FAO and OECD conclude that while additional agricultural land is available to meet future food and biofuel demand, economic and environmental factors may limit agricultural development.[25] In this context, IAASTD[4] called for a change in approach if agriculture is to meet the needs of the developing world, especially in the face of uncertainties regarding energy, competition for natural resources, climate change and threats to the natural environment. The assessment promoted the concept of 'multifunctional' agriculture and land use to provide a range of benefits simultaneously in order to enhance human well-being. It explored how traditional knowledge, science and technology could be harnessed to alleviate poverty, achieve food security, and improve environmental sustainability – essentially improving the productivity of farming systems, in the broadest sense.

The concept of sustainable agriculture has been further elaborated by the UK's Royal Society in a review of the contribution of science and the sustainable intensification of global agriculture.[14] It argues that the needs of the growing population must and can be met from the existing global agricultural land if irreversible damage to the world's ecosystems is to be avoided. The review calls for a large-scale 'sustainable intensification' of global agriculture in which yields are assessed not just per hectare, but also per unit of non-renewable inputs, especially fossil fuel, and impacts upon ecosystem services. The review identifies a key role for agricultural science including new crop varieties and appropriate agro-ecological crop and soil management practices. It argues for an inclusive approach that develops capacity in the governance of sustainable agricultural systems.[26]

These arguments are further reinforced by the *Foresight Review of the Future of Food and Farming*[5] which concluded that if: (i) there is relatively little new land for agriculture, and (ii) more food needs to be produced, and (iii) achieving sustainability is critical, then sustainable intensification is a priority. The Commission on Sustainable Agriculture and Climate Change also concludes that a 'business as usual' approach will not deliver global food security, calling for a commitment to support science-based strategies to improve agricultural production mainly from the existing agricultural land areas whilst operating within safe environmental limits.[27]

5   Case Study: UK Agriculture and Land Use

The relationship between modern agriculture and land use can be explored at the country level using the case of the UK. This assessment draws on work carried out for the *Foresight Land Use Futures Project* that took a strategic view of land resources in the 21st century.[5] Agricultural land in the UK plays diverse roles: providing food and contributing to the economy, and increasingly, to wider environmental agendas. Agriculture and agricultural land use have been strongly influenced by UK and European Government policy which in turn reflect changing circumstances and priorities. Four phases are apparent:

  i. *Growth:* beginning with the 1947 Agricultural Act, promoting self-sufficiency through guaranteed minimum prices and incomes, producer marketing boards and farm development schemes;
 ii. *Consolidation:* from joining the European Economic Community in 1972, characterised by protectionism and production support under the Common Agricultural Policy (CAP) regime;
iii. *Adjustment:* from the mid 1980s, involving production quotas, payments per head of livestock and per hectare of crop, agri-environment schemes, growing concern about environmental impacts of agriculture and introduction of EU environmental directives;
 iv. *Reform:* since the early 2000s, involving 'decoupling' of farm income support and commodity subsidies, increased environmental regulation and stewardship, the Single Payment Scheme and compliance with good farming practice. CAP reform after 2013[28] seeks to achieve food security, a better environment and a living countryside, while meeting the challenge of climate change.

These policy phases, along with other factors such as changes in domestic and world markets, technology, institutional and organisational arrangements, and in the motivation and behaviour of farmers, have acted as drivers of change in the farming sector. Their impact is evident in changing land use, employment, farm size and farming practices. At times, this dynamic has been further influenced by unforeseen events such as the outbreak of BSE in the early 1990s and Foot and Mouth epidemics in 1968 and 2000. Very recently, following a period of relative decline in farming economics, the global food shortages of 2007–2008 led to unprecedented peacetime increases in food prices, a rapid positive response from UK farmers, and a reminder of the strategic importance of agriculture and its role in food security.

5.1   Trends in UK Agriculture

A brief review of trends in UK agriculture shows how the sector has responded to changing drivers in the past and how it might in future. There are important regional differences: the north and west of the UK is dominated by grassland

and livestock, and the south and east by arable systems.[29] As farming has become more specialized, with less integration of livestock and arable farming, regional differences in land use and farming systems are more evident. As a result, the impact of changes in agricultural markets and policies can be very different between regions, as indeed can be the effects of climate change. Indeed, diversification of agricultural enterprises and land use can enhance ability to adapt to uncertain climatic conditions.

5.1.1 Agriculture's Contribution to UK Economy.   Generally, in line with other EU and OECD member states, agriculture's share of UK national economic output and employment has declined over time as overall prosperity has increased. These declined from 5% and 6% of GDP and employment respectively in the 1950s to 0.8% and 1.8% in 2007, currently providing about 530 000 full time equivalent jobs. As a comparison, in 2002 agriculture's share of GDP in the EU-15 and the USA economy was 1.7 and 1.4 per cent respectively.[30] A broader view to include the food industry and rural tourism, to which agriculture is closely linked, increases the current share of GDP to about 8% and 12.5% respectively.

5.1.2 Agricultural Land Use.   Agriculture currently occupies 77% of the area of England and Wales, high by European standards. The agricultural land area has declined by about 0.2% per year over the last 50 years, about 10% in total (Figure 4).

With respect to **crop areas**, although the total area of UK cereals has declined, the area of wheat has increased, more than doubling in the case of England. Protein crops, such as beans and peas, and oilseed rape increased from very little to over 20% of the total arable area, while root crops, mainly sugar beet and potatoes have declined in area. Horticulture is now half the area it was in the 1950s. In the last decade, the areas of forage maize as a feed for

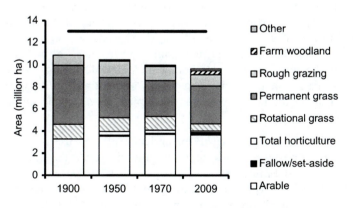

Figure 4   The area of land used for agriculture in England is decreasing (1900 to 2009).

dairy cows and farm woodland have expanded considerably. Set aside, whereby land 'surplus' to requirements are taken out of production, accounted for almost 10% of the arable area in 2007 before its discontinuation in the face of commodity shortages.

There have been considerable changes in the **livestock sector**. A combination of improved productivity, milk quotas, declining profits, and static or decreasing demand for milk, and the BSE crisis are associated with a 41% decline in the number of dairy cows in the last 30 years. Sheep and lambs numbers peaked in the 1990s in response to strong prices, with increased stocking in upland areas, but have subsequently fallen. Pig production declined mainly due to competition from imports, whereas poultry production increased in response to market growth facilitated by industry consolidation. There has been a marked decline in livestock numbers since the 'decoupling' of farm income support from livestock subsidies. The dependency of the livestock industry on government support, especially in marginal areas, is generally high, indicating a vulnerability to policy change.

These changes in crop and livestock production confirm the responsiveness of farmers to a range of policy, market and technological drivers that have influenced the relative attractiveness of particular enterprises and have generally favoured economies of scale and specialization.

5.1.3 Farm Size. Farm businesses in the UK landscape have become bigger and more specialized. The number of farms, currently about 230 000, has reduced by about 50% since 1950. Concomitantly, the average size of holdings in terms of area and livestock numbers has increased by about 40% and 150% respectively. On average, 'commercial' UK farms now are about four times larger in area than the EU average.[31] The largest farms in the UK, in terms of area and turnover, are concentrated in southern and eastern England. This consolidation has been associated with increased specialization and economies of scale associated with new technology. There has also been a recent trend towards diversification of the farm business, with 58% of farms deriving income from non-farming sources. Many farms now operate under a variety of 'partnership' arrangements, with increased use of services of agricultural contractors and advisors. Many decisions on land use are no longer made by occupying farmers only. There has, however, been an increase in the number of small hobby and lifestyle farms,[32] especially in peri-urban areas, many registering their 'agricultural' holding to claim entitlement under the new regime for 'decoupled' farm income support.

5.1.4 Farm Yields. The intensity of agricultural land use is evident in the achievement of on-farm yields of crops and livestock that have shown a continuing upward trend, although there is considerable variation amongst crop and livestock types. From the 1950s through to the mid 1980s, reductions in farmed areas were more than offset by increased yields associated with improvements in crop and livestock genetics, nutrition, and health and land improvements such as drainage.[33,34] For example, average wheat yields have doubled since the 1960s, mostly due to plant breeding technologies (Figure 5).[35,36] The importance of weather on yield is apparent from the low yields due to drought in 1976, and high yields in 1984 due to good growing conditions.

In the dairy sector, average milk yields per cow doubled to over 7500 litres between 1960 and in 2010 (Figure 6), similar to that observed in the Netherlands, Germany and France. Although yield improvement stalled following the imposition of quotas in 1984, the annual increase in productivity has been relatively consistent. Half this increase is accredited to animal breeding,[37] and the balance mainly to improved feeding regimes.[38] Thus it appears that

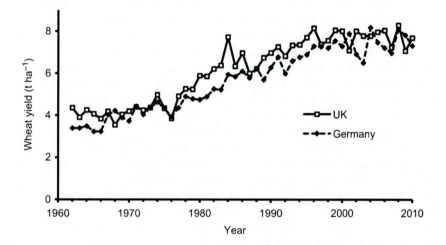

Figure 5   Annual wheat yields in the UK and Germany from 1962 to 2010.[17]

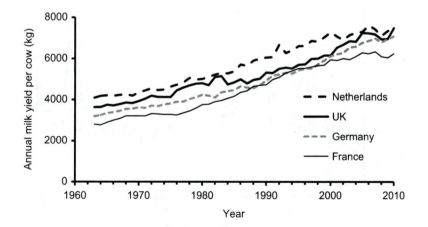

Figure 6   Observed milk yield per cow in the Netherlands, UK, Germany and France from 1962 to 2010.[17]

yield enhancing technologies, driven mainly by policy induced incentives, have virtually doubled the productive capacity of land over the last 50 years or so. Farmers, for their part, have adopted these technologies in an attempt to secure incomes in the face of declining real commodity prices and rising input costs, as explained below.

5.1.5   Agricultural Commodity Prices.   Prices for farm produce have been a main driver of land use and land management practices. World market prices for agricultural commodities have fallen for much of the last 70 years, with the exception of short lived spikes in the early 1970s, the early 1990s and, more recently, the 2007–2008 and 2010–2011 periods, mainly caused by supply disruption at a global scale.[39] Commodity prices for UK farmers have mirrored these trends especially since the early 1990s, modified by the rate of farm support within the European Union (Figure 7). While farmers have been responsive to strong prices for commodities, such as for wheat and oilseed rape in the early 1990s, they have been less responsive to weak prices. This partly reflects a degree of so-called 'asset fixity' in that farm resources, including land, have limited alternative uses, at least in the short term. This has been the experience particularly in the case of the livestock sector, especially in relatively remote, disadvantaged areas.

Cereal prices received an unprecedented boost in 2007–2008, when global deficits associated with harvest failures increased, demand for bio-fuels, and commodity speculation resulted in price spikes. Although improved supply conditions had returned prices to their mid 1990s levels by 2009,[40,41] prices increased again in 2010–2011 (Figure 7). Thus, agricultural commodity prices in the UK and hence the incentives to UK farmers largely depend on two main factors: world market prices for commodities (including oil) and the relative

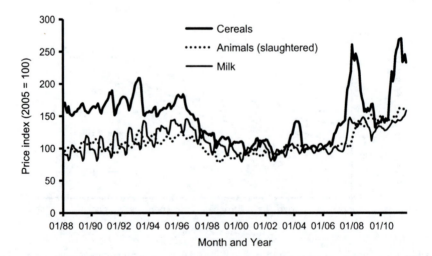

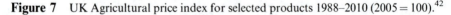

Figure 7   UK Agricultural price index for selected products 1988–2010 (2005 = 100).[42]

value of the £ to US$ and Euro exchange rates. Without a major shift in policy, UK agriculture will continue to 'take' its prices from global markets rather than exert any major influence on them. As a major global player, however, the EU as a whole has potential to influence world market prices.

5.1.6 UK Farm Incomes. The profitability of farming, a key factor influencing agricultural land use, has showed volatility around a generally declining trend, largely reflecting the trends in real commodity prices and changes in the £ Sterling to Euro exchange rate (Figure 8). In 2008, Total Income from Farming (TIFF) was about 40% lower in real terms than in 1973, when commodity prices where highly protected under CAP. Incomes increased from 1992 to 1996 as the relative international value of £ Sterling decreased, and then declined from 1996 to 2000 as the value of £ Sterling strengthened. It is noted that the benefits of higher output prices in 2008 were offset by sharp increases in prices for fertilisers and soil improvers.[40] The prospect of higher energy prices could challenge the viability of farming systems that depend on large inputs of inorganic fertilisers, pesticides and farm machinery.[43,44]

5.1.7 Productivity of UK Farms. Farm productivity is an important determinant of the amount of land required to satisfy the market demand for agricultural commodities. Total Factor Productivity (TFP), which measures the value weighted volume of outputs from agriculture relative to all inputs increased by about 25% from the early 1970s to the mid-1980s when production quotas were first introduced (Figure 9).

Some of the lack of productivity increase in UK agriculture between 1984 and 2000 reflects a tendency for the majority of farms to fall behind the most

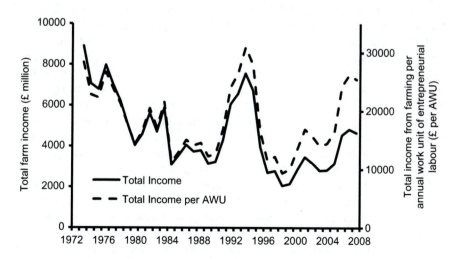

Figure 8 Total Income from Farming (TIFF) and total income from farming per annual work unit of entrepreneurial labour 1973–2010.[45]

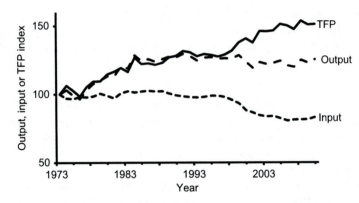

Figure 9   Levels of output index, input index and total factor productivity (solid-line: TFP) index for UK agriculture from 1973 to 2010 (1973 = 100).[46]

Table 1   Examples of the calculated agricultural total factor productivity for selected countries (TFP of USA in 1996 = 100).[47]

Year	Germany	France	Netherlands	UK
1974	48.6	47.6	78.5	50.1
1984	60.3	56.5	78.9	59.5
1994	63.3	64.4	93.4	58.1
2002	69.4	71.4	94.8	63.2

efficient, suggesting a failure to take up potential productivity gains. The improvement in TFP between 2000 and 2005 was associated with the main-tenance of a steady output but using reduced inputs such as fertiliser and labour, and by taking marginal land out of production. However between 2005 and 2010, productivity has remained relatively constant. International com-parisons of UK agriculture suggest that UK farm productivity has lagged behind that of other EU member states and USA since the mid 1980s (Table 1). TFP is an important determinant of a country's international comparative advantage and, in the event of further liberalisation of agricultural trade in future, a key determinant of whether a country imports or exports. Further-more, improved productivity in farming is now more closely attuned to improved environmental performance associated with reduced waste, precision application of chemicals, and energy and water use efficiency.[33]

Productivity gains in labour use have been associated with major changes in the landscape. Perhaps the biggest increase in productivity in UK agriculture has been in output per worker which has increased by a factor of six since the 1950s.[48,49] The continual release of labour from commercial farming, because of its expense or because it has proved difficult to recruit, has been a char-acteristic of UK agriculture, in common with the rest of Europe. The increased size of tractors and equipment has been associated with larger field sizes, removal of boundary features such as hedgerows, and installation of field drainage in order to extend the period when fields can be worked by heavy machines. Farm mechanization is in many ways the key 'agent' by which farmers engage with the land, often employing contractors to use the latest and largest equipment. Future trends in farm mechanization are a critical factor in land management, and currently a relatively under researched topic.

5.1.8 Demand for Food in the UK. Food demand depends on a combination of demographic, social and economic factors, associated with the size and make-up of the population, economic prosperity and the distribution of incomes, food preferences and spending habits.[29] Increased average income, together with increased food choices and changing tastes and lifestyles, have been associated with changes in the national diet, with consequences for farming and land use, some of it in the changes in crop and livestock production referred to above. For example, the per capita consumption of milk and dairy products, bread and fresh potatoes has fallen significantly in recent years, while the consumption of fruit and vegetables has increased. Although total meat consumption has remained relatively steady, there have been large fluctuations in the type of meat bought, partly due to food safety concerns, such as BSE, and partly due to concerns about health and diet.[50] Poultry consumption now exceeds that of other meats in the UK.[50]

In recent years, the demand for organic and 'welfare' foods has increased substantially, and concern about excessive food miles has promoted interest in locally procured seasonal foods.[51,52] However, consumers identify cost and accessibility as major barriers to consuming seasonal and local foods.[53] Demand for these foods faltered in 2007 following the onset of the economic recession, demonstrating the relative sensitivity of their demand to changes in disposable income. It is unclear as yet how these trends affect aggregate land use, but organic and animal welfare systems require additional land use compared with conventional agricultural production systems because of relatively lower yields.[54]

Coinciding with these market changes, market power has consolidated in the food retail sector in the last two decades[4,55] such that about 75% of food sales are now made by 'supermarkets'. This concentration has tended to favour larger farm producers.[4,56,57] The supermarkets have, however, promoted improved environmental and welfare performance through produce assurance schemes, in parallel with farmer environmental auditing schemes such as Linking Environment and Farming.[58]

The UK Government's vision for Food 2030[59] seeks to enable and encourage people to eat a healthy, sustainable diet. This includes wasting less food, eating food that is in season, and buying foods that are produced sustainably. These aspects have implications for land use and management.

5.1.9 Food Self-sufficiency and Food Security. The amount of land used by agriculture reflects the degree of self sufficiency in foods that can be produced in the UK. This remained at about 75% of UK requirements for indigenous foods in the mid 1970s when 'Food from Our Own Resources' was the main policy objective.[60] It reached a peak in 1984 when self sufficiency reached 78% for all foods and 95% for indigenous food that can be grown in the UK. However by 2009, the level of self-sufficiency had declined to 59% for all food and 73% for indigenous food.[61] Much of this decline is a result of reduced beef production and increased imports of pork and milk products. About two thirds of imported food now comes from the EU.[61] Thus, as previously mentioned, the incentives and rewards for UK farmers, and the land use and farming systems that result, are increasingly a product of international commodity markets, exchange rates, and the agricultural and food policies of major exporting and importing countries and regions, including the EU.

The UK Government does not prescribe particular levels of national self sufficiency. Rather it refers use 'food security' to denote guaranteed household access to affordable nutritious food.[59] UK agriculture, along with the food industry as a whole, is charged with 'ensuring food security through a strong UK agriculture and international trade links with EU and global partners which support developing economies'.[59] In this regard, UK agriculture is required to be internationally competitive.

6 Agriculture and Environment

Agriculture's role extends well beyond the production of food and fibre to include the provision of a wide range of 'public goods' associated with managed landscapes and habitats, public access to the countryside, and the regulation of water and atmospheric gases. It also produces a number of 'public bads' associated, for example, with soil degradation, water pollution, habitat loss, and greenhouse gas emissions (Table 2). The Environmental Accounts for Agriculture[62] gives some, albeit incomplete, estimates of the positive and negative environmental externalities associated with agricultural land use. Until 2008, these were revised and extended by Defra,[63] suggesting that the net environmental costs of agriculture in the UK have been decreasing in real terms since 2000.

In 2007, the estimated environmental externalities of UK agriculture amounted to a net cost of about £830 million per year (equivalent to £14 per head of population). Gross annual environmental benefits were about £1.74 billion (about £29 per head of population), mainly associated with agriculturally managed landscapes and habitats. This probably underestimates the real value of managed landscapes, especially for tourism and recreation: the impact of the travel restrictions on tourism and the rural economy due to the 2001 Foot and Mouth epidemic, for example, was estimated at £5 billion.[64]

Gross annual environmental costs were about £2.57 billion in 2007 (about £45 per head), mainly associated with soil- and livestock-related emissions to air at about £2 billion and water-related damages at about £0.5 billion. These estimates need cautious interpretation, being based on many assumptions and a somewhat unrealistic comparison of the situation 'with' and 'without' agriculture in the UK. Estimates of the extra environmental benefits and cost of

Table 2 Summary of environmental accounts for UK agriculture, 2007.[62]

	Externality	*Annual flow (£ million)*
Benefits	Landscape	616
	Biodiversity	1088
	Waste services	37
	Total	1741
Costs excluding emissions to air	Flooding	244
	Fresh water	144
	Drinking water	160
	Soil erosion	11
	Waste	7
Net benefits	Total costs	566
	(excl. emissions to air)	1175
Costs from emissions to air	Climate change	1371
	Air quality	634
	Total emissions to air	2005
Total benefits less costs		(830)

marginal changes in agricultural land use and practices would be more useful. These figures are, however, indicative of substantial positive and negative externalities that should be accounted for in decisions that concern agricultural development.[65] The figures can also be compared with farming's net value added at market prices of about £2.9 billion given in 'Agriculture in the UK' and gross value added of £5.7 billion.[40]

As a result of the importance of agriculture in the landscape, the relationship between farming and land has become the subject of an increasingly comprehensive range of policies, many of them contained in European Directives such as *The Habitats Directive*, *The Nitrates Directive* and *The Water Framework Directive*. Within agricultural policy itself, a suite of agri-environment schemes now pay famers to protect the natural environment and provide environmental improvements. There are currently, for example, over 6 million hectares of farm land under agri-environment and related schemes in England.[66] Additionally, in return for income support, farmers are required to keep their land in 'good agricultural and environmental condition'. It remains to be seen how these schemes will develop in future, especially after the 2013 CAP reforms which are likely to make a bigger distinction between policies to protect and enhance the rural environment and those that enable agriculture to compete effectively in global markets, whether producing for domestic consumption or export.

7 Agriculture and Ecosystem Services

It is apparent from the above that land can support multiple activities within the same space, providing a range of different goods and services simultaneously, such as farming and nature conservation, and farming and carbon sequestration. This is often referred to as 'multifunctionality', whereby land produces 'multiple outputs and, by virtue of this, may contribute to several societal objectives at once.'[67] Modern trends in agricultural land ownership and management, however, have tended to promote a single main purpose to the exclusion of others, often in response to dominant market or policy drivers. Production-oriented farm subsidies, for example, have encouraged intensive agriculture often at the expense of biodiversity. As a result many farmed landscapes have become less varied in appearance, in the services they provide and in the values they generate.

Multifunctional landscapes can create more value for people and communities than single-function land uses. They are also likely to be more resilient and sustainable in the long term.[68] The challenge is to recognise the value of multiple flows of services and build these into market or policy instruments that reward the provision of services, or penalise their loss.[69]

In the last decade or so, the concept of 'ecosystem services' has been developed to show the links between the 'health' of major ecosystems such as uplands, 'enclosed' farmland and floodplains and the generation of diverse flows of services that enhance the well-being of people and communities.[70-73] There are different ways of categorising the benefits provided by ecosystems [74-76] and the consequences of their degradation. Table 3 shows the framework used in the recent UK National Ecosystem Assessment (UKNEA).[72] Agricultural land can generate a range of ecosystem services, particularly regarding the 'provisioning' of food, fibre and bio-energy. Agriculture also impacts on 'regulating' services, such as flood control and water quality, as well as the cultural services relating to landscape and biodiversity. Furthermore, the management of agricultural land has potential to affect fundamental processes, such as soil formation and nutrient cycling that 'support' the other services which directly benefit people.

The UKNEA, in its review of the services generated by enclosed farmland, concluded that, although the situation has improved since the 1990s, the

Table 3 *The UK National Ecosystem Assessment*[72] *categorised the services provided by ecosystems, such as agricultural land, in terms of provisioning, regulating and cultural services which are underpinned by supporting services.*

Ecosystem services		
Provisioning services	*Regulating services*	*Cultural services*
Including crops, livestock, fish, trees, standing vegetation, water supply and wild species diversity	Including climate regulation, pollination, pollution control, hazard regulation, noise regulation, and disease and pest regulation	Including wild species diversity, and environmental settings in terms of landscape, recreational, spiritual and heritage opportunities
Supporting services		
Primary production, Soil formation, Water and Nutrient cycling		

provisioning of agricultural products continues to have a negative effect on regulating services linked to water quality, carbon emissions, and flood runoff from farm land, as well as habitat loss.[77] Consistent with other reviews,[4,5] UKNEA identifies a need to enable agriculture to enhance other ecosystem services while continuing to increase food production. This theme also features strongly in recent policy statements for food[59] and nature conservation.[78]

In this context, there is growing interest in 'payments for environmental services' (PES) as a means of converting non-market values of the environment into real incentives for land managers.[79,80] PES involve voluntary transactions to exchange well-defined environmental services between service buyers and service sellers. Most PES schemes operate through specific land uses capable of producing the required environmental service, such as field margins, woodlands or wetlands, rather than focusing on specific outcomes such as wildlife numbers or carbon flows that are more difficult to measure.

While some PES schemes are financed by users for commercial benefit, such as water companies wishing to secure water supply and quality,[81] most are funded by governments providing public goods such as biodiversity, flood control and public access in the countryside.[66] About 66% of utilisable agricultural land in England is now registered under PES-type Stewardship Schemes, with about 50% of farm land under the Entry Level Scheme. Take-up of the Higher Level Scheme (6% of the agricultural area) has been limited given the considerable changes in land use required and the administrative burdens involved.[66] Future take-up will be dependent on the strength of farm commodity prices and the relative profitability of commercial farming.

8 Agriculture and Climate Change

Agriculture has an important role in the mitigation of and adaptation to climate change.[82] It can reduce its emissions by changes in farming practices. Agricultural land can sequester and store carbon, as well as produce substitute products such as biofuels that can have a lower carbon footprint than fossil fuels. It can help to retain water in the landscapes in order to mitigate flood risk to urban areas.[83] Farmers will need to adapt to changes in weather patterns, water availability and possible pests and disease problems that may be associated with climate change.

Agriculture accounts for about 7% (51 Mt CO_2 equivalent) of total annual greenhouse gas emissions (GHG) in the UK,[84] associated with fossil fuel usage (11% of agricultural GHG), methane from livestock (37%), and nitrous oxide from fertiliser use (52%). Release of soil carbon is currently low,[85] but degradation of soils could significantly add to annual emissions; UK soils store the equivalent of 50 times the total UK annual emission of GHG. The food system as a whole, including processing and distribution, accounts for about 18%–20% of total UK GHG emissions,[86,87] about half of which is attributable to food production, allowing for emissions associated with imported foods.[88]

The complex biological processes involved in crop and livestock production make it difficult to reduce GHG emissions compared with other sectors,

especially of nitrous oxide and methane.[84,88–90] Although there are no specific
emission targets for the agriculture, there is scope to improve the efficiency of
fertiliser use, livestock feeding and breeding, waste management and cultivation
practices.[91–93] The UK Government's low carbon transition plan anticipates a
relatively modest 6% reduction in total agricultural emissions on 2008 levels by
2050, supported by technical assistance and bio-energy capital grants.[84] Other
measures, such as reduced consumption and production of livestock products
associated with changing human diets could achieve further reductions of
15–20% in GHG emissions from agriculture by 2050.[44]

Agricultural land also offers potential to offset the emissions from other
sectors through carbon absorption and storage.[33,82,92] In future, this could
feature as a part of emission trading schemes, linked to the provision of and
payment for other services such as nature and natural resource conservation.

9 Agriculture and Energy

UK agriculture, although accounting for only about 2% of national energy
consumption, is vulnerable to the likelihood of higher and more volatile oil
prices. Much of the gain in agricultural productivity in the UK in the last 50
years have depended on relatively cheap and secure energy supplies. Although
in recent years higher oil prices have been associated with higher agricultural
commodity prices (Figure 10), potential benefits to UK farmers are offset by
increased energy costs.[43] High energy costs could encourage less intensive
farming with lower yields per hectare, increasing the land area required for
agriculture. Although organic production uses less fertilisers and chemical
inputs, it tends to use more mechanised power and has lower yields than
conventional farming, resulting in similar overall energy efficiencies.[4,54,88]
Higher energy prices will, however, induce energy saving technologies,
including on-farm energy recovery from wastes.[33]

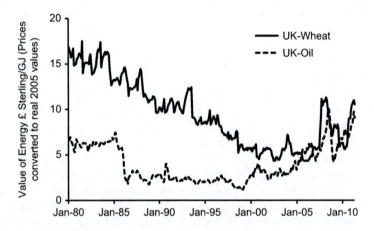

Figure 10   Converging agricultural and energy prices – real price of wheat and oil
expressed in £ Sterling per Gigajoule of energy 1980–2011.[94]

Agriculture can of course produce renewable energy from biomass. Rising oil prices, especially beyond about £5 per gigajoule of energy, mean that the cost of a unit of energy in oil approaches the cost of the energy in agricultural commodities such as wheat (Figure 10). A review for UK Government[95] concluded that a 'sustainable' biofuels industry must 'avoid using agricultural land that would otherwise be used for food production'. Greater demand for biofuels could lead to displacement of existing agricultural production,[24] increased land use change, loss of biodiversity, pressure on water resources and, as a consequence, make little contribution to reducing GHG emissions.

For the UK, using oilseed rape to meet the aspirational EU 10% target for the proportion of road transport fuels to be met by renewable sources would require the diversion of the existing oilseed rape crop of 600 000 ha and an additional 840 000 ha, equivalent to about 30% of the existing arable area.[29] This is a similar proportion to that used to feed horses at the end of the 19th century. Using a sugar beet feedstock would require about 10% of UK arable land. *The Gallagher Review* recommended that targets for transport fuels higher than 5% should only be adopted if biofuels can be shown to be sustainable, especially regarding impacts of land use change.[95]

Attention is switching from conventional crops for biofuels to second generation biomass crops, such as *Miscanthus* and willow,[96] much of which can be grown on poorer land, within the limits of available water. Third generation biofuels, especially algae-based biodiesel,[4] are now being developed, but these also may require large-scale commitments of land and water resources.[4] The increase in potential to focus on the bio-fuel production from land means that the traditional divide between agriculture and forestry becomes blurred, and hence there will be an increased need for land use decision to encompass both considerations of agriculture and forestry. It appears that there is probably sufficient land in the UK to maintain current levels of self sufficiency in food and go some way towards meeting EU targets for transport biofuels, but this will require improvements in productivity as well as adequate financial incentives to biofuel producers, whether driven by oil prices or policy measures.

10 Future Prospects for Land Use in the UK

Future changes in the scale and intensity of agricultural land use in the UK will be influenced by the interaction of factors which shape the demand for and supply of agricultural goods and services.

On the demand side, much depends on the characteristics of the markets for agricultural products, shaped by demographic, economic and social factors. These will include not only demand for conventional traded commodities, increasingly set in a global context, but also non-market environmental goods and services such as landscape management and public access to the countryside at the local scale and services such as carbon sequestration at the global scale.

On the supply side, much depends on the development and adoption of agricultural technologies that can improve agricultural productivity and its environmental performance. As we have seen, agricultural land use will also be shaped by future agricultural policies that determine the relationship between UK and world agriculture, particularly the extent to which the UK seeks to meet its own food requirements and/or engages in international trade. Agricultural land use will also be also strongly influenced by future environmental policies that determine the relationship between UK agriculture, natural resources and the wider environment. The need to mitigate and adapt to the effects of climate change will have implications for the role of agriculture and land use.

10.1   Scope to Release Land from Agricultural Production

Many of the technological innovations described above will increase land productivity and thereby determine the amount of land that is needed to meet the national demand for agricultural products. They will also determine the proportion of land that can be released from agricultural production for other purposes, whether urban development or environmental services such as habitat conservation and opportunities for recreation and carbon sequestration.

Between 1992 and 2006, the reduced need to allocate land to crop production at a European level allowed between 5 and 10% of the arable land to be 'set aside'. The European Union has been able to draw on this land bank in response to commodity shortages since 2007–08. During the next 50 years, continued improvements in technology will be pivotal in providing flexibility in future land use decisions in the face of increased demands for crop and animal products. However, the amount of land released in the UK will also depend on domestic market conditions, government interventions and international trade.

By way of example, in 2005, Morris *et al.*[97] explored possible agricultural futures and the implications for the environment for England and Wales. Table 4 shows the possible effect of 'business as usual' and four other intervention scenarios on the change in technical efficiency, self-sufficiency and the use of lowland agricultural land in England and Wales for agriculture, in 2050 relative to 2002. There is little or no surplus land for scenarios with relatively lower technical efficiency and associated yields. The 'local stewardship' scenario also requires relatively high agricultural land use, including in the uplands to meet self-sufficiency targets. Crop prices tend to be higher when biofuels are produced to meet EU targets and land use greater than otherwise.

Scenarios have also been used at the regional UK and EU scale to explore land use change[82,98–101] allowing for changes in demand for agricultural commodities, agricultural productivity and the effects of climate change. Biofuels were shown to substitute for food production, especially if energy prices and yields of energy crop increase. At the EU level, increases in productivity were predicted to exceed growth in demand such that large areas of land would be taken out of agricultural production, with surplus land moving into urban, forestry and recreational uses. This appears consistent with past land use

Table 4 Predicted effect of 'Business as Usual' and four other intervention scenarios on the change in technical efficiency, self-sufficiency and the use of lowland agricultural land in England and Wales for agriculture, in 2050 relative to 2000–2004 values (derived from Morris *et al.*, 2005).[33,97]

Scenario	Agricultural intervention regime	Proportional change in technical efficiency (%)[a]	Proportional change in self-sufficiency (%)[b]	Proportional change in land use for agriculture (%)
Business as usual	Agricultural support as in 2002	+19	+6	−20
World markets	Zero: market-driven free trade	+34	−3	−34
National enterprise	Moderate/high: protected domestic markets with limited environmental concern	+39	+26	−18
Global sustainability	Low: market orientation with targeted sustainability compliance	+12	+8	−2
Local stewardship	High: locally defined schemes reflecting local priorities	−7	+23	0

[a]Mean yield per hectare (or yield per head) for five crop and five livestock commodities.
[b]Expressed as the ratio between production and consumption of twelve commodities.

trends, where, even with subsidies, the area of agricultural land in Europe has decreased by 13% since the 1950s.

Other EU-scale scenario studies,[102,103] predict that the area of arable land would decrease under liberalised world markets compared with continued support afforded by the CAP regime. One study[104] predicts a high level of abandonment of agricultural land for 2030, ranging from 2.5 to 13 per cent of the agricultural areas of the EU-15 in 2000, highest where agriculture is exposed to a liberalised world market and where there is pressure for urban growth.

The recent UKNEA report[72] uses expert judgements supported by modelling to develop six scenarios that reflect differences in four main drivers: population pressure, global economic factors, technology and societal preferences. These are: (i) 'Green and Pleasant Land', where a priority is given to the conservation of ecosystems; (ii) 'Nature@Work' where ecosystem services are promoted through the creation of multifunctional landscapes; (iii) 'Local Stewardship' where society strives to be sustainable within its immediate surroundings; (iv) 'Go with the Flow' where current trends are assumed to continue, (v) 'National Security' where there is reliance on greater self-sufficiency in food and resources; and

(vi) 'World Markets' where the goal is economic growth and the elimination of trade barriers.

Enclosed farmland remains the dominant land use under all scenarios, increasing under 'National Security' and decreasing under 'World Markets'. Measures taken to conserve a range of ecosystem services under 'Green and Pleasant Land' and 'Local Stewardship' are associated with low-input output farming and, to varying degrees, reductions in farmland areas released for other uses. 'Nature@Work' and 'Go with the Flow' assume the improvement of productivity due to more intensive yet sustainable farming methods.

Most predictions point towards a decline in land use for agriculture in the UK under a 'business as usual' scenario. Agricultural land for bio-energy production will be an important factor in determining this decline and its influence on prices of other crops. The scenarios also point, in general, towards an increase in productivity and efficiency of farming methods. The influence of world market prices for agricultural commodities and the direction of CAP reform are common themes affecting land use futures. In general socio-economic, technological and political factors are considered likely to be the most important drivers for land use change in the next 30–40 years.[82]

10.2   Technology Change and Land Use

Regarding the role of agricultural science and technology as it affects land use, Figure 11 develops the systems framework shown earlier in Figure 1 to demonstrate how seven key aspects of technological change can help improve agriculture productivity and environmental performance with implications for future land use.[33]

The first three aspects (Figure 11, Table 5) mainly concern improvement in inputs to farming systems, namely genetic material, reducing abiotic and biotic

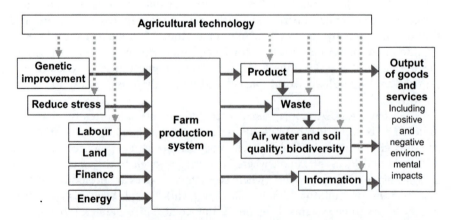

Figure 11   Schematic diagram showing seven key areas (indicated by the vertical dotted lines) where technology can affect the performance of agricultural systems and thereby land use.

Table 5 Seven areas within agricultural systems which can be affected by technological change, example technologies, typical aims in using the technology and possible effects on land use.[33]

Area	Example technologies	Typical aim in using the technology	Effect on land use
1. Genetic improvement	Plant and animal breeding, and genetic modification.	To increase yield and/or quality per area of crop or per animal, or per unit time.	Often incremental, but product changes can result in step-changes.
2. Reducing abiotic and biotic stress	Crop nutrition and protection; irrigation, drainage; animal nutrition and health; and housing and habitat management.	To minimize yield and quality losses due to stress and injury; and to improve animal welfare.	Usually incremental.
3. Labour and energy productivity	Mechanisation and herbicides.	To increase output per unit labour and energy and improve the timeliness of operations.	Usually incremental.
4. New agricultural products	Biomass for energy, bioethanol, biodiesel and biogas; Mediterranean crops; health-based products; and change in product mix.	To develop new products or products mixes in response to new markets or environmental change.	Often a step change.
5. Reducing waste	Waste as energy; waste as feedstocks; plant and animal breeding; and food storage.	To minimise waste but to maximise its value.	Usually incremental, but regulation can cause step-changes.
6. Enhancing air, soil and water quality and biodiversity	Minimum tillage; precision farming; animal nutrition; habitat management; and catchment-sensitive farming.	To reduce net greenhouse gas emissions and to enhance air, water and soil quality and farm biodiversity.	Usually incremental, but regulation can cause step changes.
7. Information transfer and use	Computerisation and biosensors.	To improve decision making by farmers and consumers.	Usually incremental.

stress, and increasing labour and energy productivity. In the main, the impact on land use and land management is incremental as farmers adopt new technologies over time and in response to market incentives or pressures to reduce costs. A review of yields of farmed species into the 21st century[34] identified scope for increases of 'average farm' yields of up to 30%–60% by 2050 depending breeding technologies and farmer take-up. The use of genetically modified (GM) organisms is perceived by some to offer potential benefit to reduce susceptibility to yield losses, particularly associated with disease or pest problems. While GM technologies are not perceived to offer substantial yield gains compared with conventional breeding technologies, they could enable yield enhancing traits to be derived more quickly.

The fourth aspect involves new agricultural products associated for example with bio-energy and other industrial crops, and specialist niche crop and livestock products for new and emerging markets linked for health related or ethical foods. Such developments could involve step changes in land use and farming practices, similar in some respects to the introduction of oilseed rape into the UK landscape in the 1970s, and maize for animal feed more recently.

The last three aspects focus mainly on the management of outputs such waste production, the effect on air, water and soil health and biodiversity, and information transfer. These technologies are both remedial, helping to reduce the environmental burden of agriculture, and potentially enhancing, making better use of available resources. They are likely to be driven by increased environmental regulation and market requirements to demonstrate compliance with good environmental practice. In future, agricultural waste will be seen as a resource rather than something to be disposed of. Small scale anaerobic digestion of animal wastes for example can simultaneously provide energy and alleviate the environmental burden. Reduction in waste throughout the food chain could significantly reduce the demand for raw food, possibly by between 10 and 20%.

There is a clear role for technologies which control agricultural impacts on air, water, soil and biodiversity, including minimum tillage, precision application of chemical inputs, greater use of natural predators in pest and disease control, and possibly genetic engineering. The use of nano-technologies for example offers considerable scope for monitoring soil condition and releasing ameliorative treatments.[105,106] These technologies will help agriculture to operate within its environmental limits, as well as cope with the increased pressure on natural resources induced by climate change. It is likely that farming, like other industries, will be increasingly required to conform with prescriptions on best practices, such as cultivation methods, animal welfare conditions, and chemical use, especially in vulnerable or ecological sensitive areas.

Over the next 50 years, measures to reduce net carbon dioxide, nitrous oxide and methane emissions from agriculture will be particularly critical,[107] especially through soil and livestock management. Finally, information transfer and use are likely to continue to be a key area of technological innovation.

Innovations such as the internet can provide the farmer with a greater understanding of consumer requirements. Computer aided decision support tools can help inform the use of organic and chemical fertilizers, crop protection measures, animal feeding regimes and control environmental conditions in livestock housing. For their part, consumers will also become better informed of the environmental impact and energy burden of different products or different agricultural systems, such as conventional and organic agriculture,[88,107,108] and food commodities will be more readily traceable.

11 Implications for Policy

Although overall global agricultural production has more than kept pace with increased population, predicted increases in population and pressure on natural resources associated with economic development and climate change, will require agriculture to produce more food, more sustainably in future.

The emerging consensus is that agriculture must meet future needs for food, fibre and bio-energy mainly from the existing stock of agricultural land. Otherwise the world's ecosystems could be irreparably damaged. In this respect, it is crucial to refine existing and develop new agricultural technologies in order to improve the future sustainability of land management, as well as the knowledge and skills of land managers. Research to develop new agricultural and engineering technologies, and training and advisory services to support their adoption are critical components of strategies to develop modern agriculture, especially targeting small famers in the developing world. It will also be important to build and maintain capacity in critical agricultural assets, such as high-quality farmland, and supporting infrastructure, such as land drainage and irrigation.

For many regions of the world, the focus will be on agriculture's ability to alleviate hunger and support rural livelihoods. At the same time, however, there will be greater need to promote and reward the multiple roles of agriculture, not only as a producer of food, but also as a provider of key ecosystem services associated, for example, with climate change mitigation, protection of natural resources, flood risk management, biodiversity, and access to green space. The efficient and sustainable management of agricultural land is essential for a sustainable global society.

References

1. C. R. W. Spedding, *An Introduction to Agricultural Systems*, Elsevier, London, 2nd edn, 1988.
2. Foresight Land Use Futures Project, *Land Use Futures: Making the Most of Land in the 21st Century*, The Government Office for Science, London, 2010.
3. M. Winter and M. Lobley (ed), *What is Land For? The Food, Fuel and Climate Change Debate*, Earthscan, London, 2009.

4. IAASTD, *Agriculture at a Crossroads. Global Report*, Island Press, Washington, DC, 2009.
5. Foresight Future of Food & Farming Project, *The Future of Food & Farming: Challenges and Choices for Global Sustainability*, The Government Office for Science, London, 2011.
6. United Nations, *World Population Prospects: the 2010 Revision: Highlights and Advance Tables*, ESA/P/WP.220, Population Division of the Department of Economic and Social Affairs of the United Nations Secretariat, New York, 2011.
7. Food and Agriculture Organization, *World Agriculture towards 2030/50: Prospects for Food, Nutrition, Agriculture & Major Commodity Groups. Interim Report*, Rome, 2006.
8. Foresight Tackling Obesities Project, *Tackling Obesities: Future Choice*, Government Office for Science, London, 2007.
9. R. J. Hodges, J. C. Buzby and B. Bennet, Postharvest losses and waste in developed and less developed countries: opportunities to improve resource use, *J. Agric. Sci.*, 2011, **149**, 37–45.
10. J. Gustavsson, C. Cederberg, U. Sonesson, R. van Otterdijk and A. Meybeack, *Global Food Losses and Food Waste*, Food and Agriculture Organization, Rome, 2011.
11. Food and Agriculture Organization, *The State of Food Insecurity in the World 2011*, Rome, 2011.
12. US Energy Information Administration, *International Energy Outlook 2011*, Washington, DC, 2011.
13. J. Woods, A. Williams, J. K. Hughes, M. Black and R. Murphy, Energy and the food system, *Philos. Trans. R. Soc. London, Ser. B*, 2010, **365**, 2991–3006.
14. The Royal Society, *Reaping the Benefits. Science and the Sustainable Intensification of Global Agriculture*, London, 2009.
15. United Nations Convention to Combat Desertification, *Desertification: A Visual Synthesis*, UNCCD Secretariat, Bonn, Germany, 2011.
16. R. Lesslie, J. Mewett and J. Walcott, *Landscapes in Transition: Tracking Land Use Change in Australia*, Australian Bureau of Agricultural and Resource Economics and Sciences, Canberra, Australia, 2011.
17. Food and Agriculture Organization, *FAOSTAT*, 2011; http://faostat. fao.org/site/291/default.aspx (accessed January 7, 2012).
18. S. Pfister, A. Koehler and S. Hellweg, Assessing the environmental impacts of freshwater consumption in LCA, *Environ. Sci. Technol.*, 2009, **43**(11), 4098–4104.
19. International Water Management Institute, *Water for Food, Water for Life: Comprehensive Assessment of Water Management in Agriculture*, International Water Management Institute and Earthscan, London and Colombo, Sri Lanka, 2007.
20. J. Gornall, R. Betts, E. Burke, R. Clark, J. Camp, K. Willett and A. Wiltshire, Implication of climate change for agricultural productivity in

the early twenty-first century, *Philos. Trans. R. Soc. London, Ser. B*, 2010, **365**, 2973–2989.

21. International Food Policy Research Institute, *High Food Prices: The What, Who and How of Proposed Policy Actions, Policy Brief*, Washington, DC, 2008.

22. World Bank, *Rising Food Prices: Policy Options and the World Bank Response*, Background Paper, 2008.

23. Department for Environment, Food and Rural Affairs, *The Impact of Biofuels on Commodity Prices*, London, 2008.

24. P. J. Burgess, M. Rivas Casado, J. Gavu, A. Mead, T. Cockerill, R. Lord, D. van der Horst and D. C. Howard, A framework for reviewing the trade-offs between renewable energy, food, feed and wood production at a local level, *Renewable Sustainable Energy Rev.*, 2012, **16**, 129–142.

25. OECD/FAO, *OECD/FAO Outlook for Agriculture 2009–2018*, Organisation for Economic Cooperation and Development and the Food and Agriculture Organization of the United Nations, Paris and Rome, 2009.

26. M. Pimbert, *Participatory Research and On Farm Management of Agricultural Biodiversity in Europe*, International Institute for Environment and Development, London, 2011.

27. J. Beddington, M. Asaduzzaman, A. Fernandez, M. Clark, M. Guillou, M. Jahn, L. Erda, T. Mamo, N. Van Bo, C. A. Nobre, R. Scholes, R. Sharma and J. Wakhungu, *Achieving Food Security in the Face of Climate Change: Summary for Policy Makers from the Commission on Sustainable Agriculture and Climate Change*, CGIAR Research Program on Climate Change, Agriculture and Food Security (CCAFS), Copenhagen, Denmark, 2011.

28. European Commission, *The Common Agricultural Policy After 2013: Legal Proposals*, Brussels, 2011.

29. A. Angus, P. J. Burgess, J. Morris and J. Lingard, Agriculture and land use: Demand for and supply of agricultural commodities, characteristics of the farming and food industries, and implications for land use in the UK, *Land Use Policy*, 2009, **26S**, S230–S242.

30. A. Normile and J. Price, *The United States and the European Union: A Statistical Analysis*, US Department of Agriculture, 2004.

31. Eurostat, *Key Statistics on Europe*, Luxembourg, 2009.

32. R. Munton, Rural land ownership in the United Kingdom: Changing patterns and future possibilities for land use, *Land Use Policy*, 2009, **26S**, S54–S61.

33. P. J. Burgess and J. Morris, Agricultural technology and land use futures: the UK case, *Land Use Policy*, 2009, **26S**, S222–S229.

34. R. Sylvester-Bradley and J. Wiseman, *Yields of Farmed Species, Constraints and Opportunities in the 21st Century*, Nottingham University Press, Nottingham, UK, 2005.

35. R. B. Austin, Yield of wheat in the United Kingdom: recent advances and prospects, *Crop Sci.*, 1999, **39**, 1604–1610.

36. I. MacKay, A. Horwell, J. Garner, J. White, J. McKee and H. Philpott, Reanalyses of the historic series of UK variety trials to quantify the contributions of genetic and environmental factors to trends and variability in yield over time, *Theor. Appl. Genet.*, 2011, **122**, 225–238.
37. J. E. Pryce, M. D. Royal, P. C. Garnsworth and I. L. Mao, Fertility in the high producing dairy cow, *Livestock Prod. Sci.*, 2004, **86**, 125–135.
38. M. D. Eastridge, Major advances in applied dairy cattle nutrition, *J. Dairy Sci.*, 2006, **89**, 1311–1323.
39. J. Piesse and C. Thirtle, Three bubbles and a panic: an explanatory review of recent price events, *Food Policy*, 2009, **34**, 119–129.
40. Department for Environment, Food and Rural Affairs, *Agriculture in the UK 2008*, London, 2009.
41. M. Peters, S. Langley and P. Westcott, Agricultural commodity price spikes in the 1970s and 1990s: valuable lessons for today, *Amber Waves*, 2009, **7**, 16–23.
42. Department for Environment, Food and Rural Affairs, *Agricultural Prices Indices*, 2011; http://www.defra.gov.uk/statistics/foodfarm/farmgate/agripriceindex/ (accessed 01/11).
43. Sustainable Development Commission, *$100 a Barrel Oil. Impacts on the Sustainability of Food Supplies on the UK*, London, 2007.
44. E. Audsley, M. Brander, J. Chatterton, D. Murphy-Bokern, C. Webster and A. Williams, *How Low can We Go? An Assessment of Greenhouse Gas Emissions from the UK Food System and the Scope for Reduction by 2050. Report to WWF and Food Climate Research Network*, Cranfield University, Bedfordshire, 2009.
45. Department for Environment, Food and Rural Affairs, *Agriculture in the United Kingdom. AUK datasets – Chapter 2 Farming Incomes*, 2011; http://www.defra.gov.uk/statistics/foodfarm/cross-cutting/auk/ (accessed 01/07).
46. Department for Environment, Food and Rural Affairs, DARDNI, Scottish Government and Welsh Assembly, *Agriculture in the United Kingdom 2010*, London, 2011.
47. Department for Environment, Food and Rural Affairs, *Agriculture in the UK 2007. Tables and Charts*, 2008.
48. A. Bailey, K. Balcombe, C. Thirtle and L. Jenkins, ME Estimation of input and output biases of technical and policy change in UK agriculture, *J. Agric. Econom.*, 2004, **55**, 385–400.
49. C. Thirtle and J. Holding, *Productivity in UK Agriculture: Causes and Constraints. Report to Department for Environment, Food and Rural Affairs*, Imperial College, Wye, Kent, 2003.
50. R. Foster and J. Lunn, 40th anniversary briefing paper: food availability and our changing diet, *BNF Nutr. Bull.*, 2007, **32**, 187–249.
51. Department for Environment, Food and Rural Affairs, *Food Statistics Pocketbook 2008*, 2008; http://www.defra.gov.uk/statistics/foodfarm/food/pocketstats/ (accessed 03/12).

52. National Farmers' Union, *Farming Outlook 1st Quarter*, Stoneleigh, UK, 2008.
53. Department for Environment, Food and Rural Affairs, *Attitudes and Behaviours around Sustainable Food Purchasing*, Report SERP 1011/10, Defra Food and Farming Group, York, 2011.
54. A. G. Williams, E. Audsley and D. L. Sandars, Environmental burdens of producing bread wheat, oilseed rape and potatoes in England and Wales using simulation and system modelling, *Int. J. Life Cycle Assessments*, 2010, **15**, 855–868.
55. Department for Environment, Food and Rural Affairs, *Economic Note on UK Grocery Retailing*, Defra, Food & Drink Economics Branch, London, 2006.
56. E. McCullough, P. Pingali and K. Stamoulis, *The Transformation of the Agrifood Systems: Globalisation, Supply Chains and Smallholder Farms*, FAO and Earthscan, London, 2008.
57. World Bank, *World Development Report 2008*, Washington, DC, 2007.
58. LEAF (2012), *Linking Environment and Farming*; http://www.leafmarque.com/leafuk/ (accessed 01/10).
59. HM Government, *Food 2030*, Department for Environment, Food and Rural Affairs, London, 2010.
60. Ministry of Agriculture, Fisheries and Food, *Food from Our Own Resources*, HMSO, London, 1975.
61. Department for Environment, Food and Rural Affairs, *Agriculture in the UK 2009*, London, 2010.
62. Jacobs UK Ltd, Scottish Agricultural College and Cranfield University, *Environmental Accounts for Agriculture. Project SFS0601 Final Report for Department for Environment, Food and Rural Affairs, Welsh Assembly Government; Scottish Government; Department of Agriculture and Rural Development (Northern Ireland)*, 2008.
63. Department for Environment, Food and Rural Affairs, *Environmental Accounts for Agriculture*, 2009; http://archive.defra.gov.uk/evidence/economics/foodfarm/reports/envacc/index.htm (accessed 01/09).
64. National Audit Office, *The 2001 Outbreak of Foot and Mouth Disease*, The Stationery Office, London, 2002.
65. HM Treasury, *The Blue Book*, 2008.
66. Natural England, *Agri-Environment Schemes in England 2009. A Review of Results and Effectiveness*, Peterborough, 2009.
67. Organisation for Co-operation and Development, *Multi-functionality: Towards an Analytical Framework: Agriculture and Food*, Paris, 2001.
68. J. Banks and T. Marsden, Integrating agri-environment policy, farming systems and rural development: Tir Cymen in Wales, *Sociologia Ruralis*, 2000, **40**(4), 466–480.
69. Department for Environment, Food and Rural Affairs, *Securing a Natural Healthy Environment: An Action Plan for Embedding an Ecosystems Approach*, London, 2007.

70. Millennium Ecosystem Assessment, *Ecosystems and Human Well-Being: A Synthesis*, Island Press, Washington, DC, 2005.

71. TEEB, *The Economics of Ecosystems and Biodiversity: Mainstreaming the Economics of Nature: A Synthesis of the Approach, Conclusions and Recommendations of TEEB*, Progess Press, Malta, 2010.

72. UK National Ecosystem Assessment, *The UK National Ecosystem Assessment Technical Report*, United Nations Environment Program – World Conservation Monitoring Centre, Cambridge, 2011.

73. H. Posthumus, J. R. Rouquette, J. Morris, D. J. G. Gowing and T. M. Hess, A framework for the assessment of ecosystem goods and services; a case study on lowland floodplains in England, *Ecol. Econom*, 2010, **65**, 1510–1523.

74. R. De Groot and L. Hein, Concept and valuation of landscape functions at different scales, in *Multi-functional Land Use: Meeting Future Demands for Landscape Goods and Services*, ed. U. Mander, H. Wiggering and K. Helming, Springer, Berlin, 2007, 17–36.

75. O. Agbenyega, P. J. Burgess, M. Cook and J. Morris, Application of an ecosystem function framework to perceptions of community woodlands, *Land Use Policy*, 2009, **26**, 551–557.

76. R. De Groot, B. Fisher, M. Christie, J. Aronson, L. Braat, J. Gowdy, R. Haines-Young, E. Maltby, S. Neuville, S. Polasky, R. Portela and I. Ring, Chapter 1: Integrating the ecological and economic dimensions in biodiversity and ecosystem service valuation, in *The Economics of Ecosystems and Biodiversity: The Ecological and Economic Foundations*, ed. P. Kumar, Earthscan, London and Washington, 2010, pp. 9–40.

77. L. Firbank, R. Bradbury, D. McCracken and C. Stoate, Enclosed Farmland. Chapter 7, in *UK National Ecosystem Technical Report*, ed. UK National Ecosystem Assessment, United Nations Environment Program – World Conservation Monitoring Centre, Cambridge, 2011, pp. 197–240.

78. HM Government, *The Natural Choice: Securing the Value of Nature*, CM8082, Her Majesty's Stationery Office, London, 2011.

79. S. Wunder, S. Engel and S. Pagiola, Taking stock: a comparative analysis of payments for environmental services programs in developed and developing countries, *J. Ecol. Econom.*, 2008, **65**, 834–852.

80. N. Boatman, K. Willis, S. Garrod and N. Powe, *Estimating the Wildlife and Landscape Benefits of Environmental Stewardship. Report ro Defra and Natural England*, Food and Environment Research Agency and University of Newcastle upon Tyne, Newcastle, 2010.

81. Westcountry Rivers Trust, *Upstream Thinking*, 2011; http://www.wrt.org.uk/projects/upstreamthinking/upstreamthinking.html (accessed 01/11).

82. M. D. A. Rounsevell and D. S. Reay, Land use and climate change in the UK, *Land Use Policy*, 2009, **26S**, S160–S169.

83. J. Morris, T. M. Hess and H. Posthumus, *Agriculture's Role in Flood Adaptation and Mitigation – Policy Issues and Approaches*, Background paper for Sustainable Management of Water Resources in Agriculture, Organisation for Economic Cooperation and Development, Paris, 2010.

84. HM Government, *The UK Low Carbon Transition Plan: National Strategy for Climate and Energy*, The Stationary Office, London, 2009.

85. P. M. Haygarth and K. Ritz, The future of soils and land use in the UK: Soil systems for the provision of land-based ecosystem services, *Land Use Policy*, 2009, **26S**, S187–S197.

86. Cabinet Office, *Food Matters. Towards a Strategy for the 21st Century*, The Strategy Unit, The Cabinet Office, London, 2008.

87. T. Garnett, *Cooking up a Storm. Food, Greenhouse Gas Emissions and Our Changing Climate*, The Food and Climate Research Network, London, 2008.

88. A. G. Williams, E. Audsley and D. L. Sandars, *Determining the Environmental Burdens and Resource Use in the Production of Agricultural and Horticultural Commodities. Main report. Defra Research Project IS0205*, Cranfield University, Bedfordshire, 2006.

89. Department for Environment, Food and Rural Affairs, *The UK Climate Change Programme. London*, Department for Environment, Food and Rural Affairs, London, 2008.

90. H. Steinfeld, P. Gerber, T. Wassenaar, V. Castel, M. Rosales and C. de Hann, *Livestock's Long Shadow*, Food and Agriculture Organization of the United Nations, 2006.

91. Scottish Agricultural College, *UK Marginal Abatement Costs Curves for the Agricultural and Land Use, Land Use Change and Forestry Sectors out to 2022, with Qualitative Analysis of Options to 2050. RMP4950. Final Report to the Committee on Climate Change*, Scottish Agricultural College, Edinburgh, 2008.

92. ADAS, *Analysis of Policy Instruments for Reducing Greenhouse Gas Emissions from Agriculture. Report to Defra, RMP/5142*, ADAS, Wolverhampton, 2009.

93. P. Smith, Soaking up the carbon, in *What is Land For? The Food, Fuel and Climate Change Debate*, ed. M. Winter and M. Lobley, Earthscan, London, 2009, p. 73.

94. F. Litkei, Is wheat price primarily determined by the oil price? *MSc Thesis (unpublished)*, Cranfield University, Bedfordshire, 2011.

95. Renewable Fuels Agency, *The Gallagher Review of the Indirect Effects of Biofuels Production*, Renewable Fuels Agency, West Sussex, 2008.

96. A. Karp, A. J. Haughton, D. A. Bohan, A. B. Riche, M. D. Mallott, V. E. Mallott and S. J. Clark, Perennial energy crops: implications and potential, in *What is Land For? The Food, Fuel and Climate Change Debate*, ed. M. Winter and M. Lobley, Earthscan, London, 2009, p. 47.

97. J. Morris, E. Audsley, I. A. Wright, J. McLeod, K. Pearn, A. Angus and S. Rickard, Agricultural Futures and Implications for the Environment. *Defra Research Project IS0209*, Cranfield University, Bedfordshire, 2005.

98. M. Rounsevell, F. Ewart, R. Reginster, R. Leemans and T. Carter, Future scenarios of European agricultural land use II. Projecting changes in cropland and grassland, *Agric. Ecosyst. Environ.*, 2005, **107**, 117–135.

 99. F. Ewert, M. D. A. Rounsevell, I. Reginster, M. J. Metzger and
     R. Leemans, Future scenarios of European agricultural land use.
     1. Estimating changes in crop productivity, *Agric. Ecosyst. Environ.*, 2005,
     107, 101–116.
100. G. Busch, Future European agricultural landscapes – What can we learn
     from existing quantitative land use scenario studies? *Agric., Ecosyst.
     Environ.*, 2006, **114**, 121–140.
101. M. S. Reed, K. Arblaster, C. Bullock, R. J. F. Burton, A. L. Davies, J.
     Holden, K. Hubacek, R. May, J. Mitchley, J. Morris, D. Nainggolan, C.
     Potter, C. H. Quinn, V. Swales and S. Thorp, Using scenarios to explore
     UK upland futures, *Futures*, 2009, **41**(9), 619.
102. H. Miejl, T. Rheenen, A. Tabeau and B. Eickhout, The impact of different
     policy environments on agricultural land use, *Agric. Ecosyst. Environ.*,
     2006, **114**, 21–38.
103. B. Eickhout, H. Meijl, A. Tabeau and T. Rheenen, Economic and eco-
     logical consequences for four European land use scenarios, *Land Use
     Policy*, 2007, **24**, 562–575.
104. P. Verburg, C. Schulp, N. Witte and A. Veldkamp, Downscaling of land
     use change scenarios to assess the dynamics of European landscapes,
     Agric. Ecosyst. Environ., 2006, **114**, 39–56.
105. T. Joseph and M. Morrison, *Nanotechnology in Agriculture and Food,*
     European Nano Technology Gateway, Germany, 2006.
106. *International Assessment of Agricultural Knowledge, Science and Tech-
     nology for Agricultural Development, Agriculture at a Crossroads. Vol IV.
     North America and Europe*, Island Press, Washington, DC, 2009.
107. M. J. Glendining, A. G. Dailey, A. G. Williams, F. K. van Evert, K. W. T.
     Goulding and A. P. Whitmore, Is it possible to increase the sustainability
     of arable and ruminant agriculture by reducing inputs? *Agric. Syst.*, 2009,
     99, 117–125.
108. W. Day, E. Audsley and A. R. Frost, An engineering approach to
     modelling, decision support and control for sustainable systems, *Philos.
     Trans. R. Soc. London, Ser. B*, 2008, **363**, 527–541.

Impacts of Agriculture upon Soil Quality

R. SAKRABANI,* L. K. DEEKS, M. G. KIBBLEWHITE AND K. RITZ

ABSTRACT

Agriculture covers nearly 40% of the surface area of the earth and results in 30% of global greenhouse gas emissions and 70% of global water withdrawal. Soil quality plays a fundamental role in ensuring viable and sustainable agricultural production. The impacts on soil quality associated with agriculture include a decline in soil organic matter, compaction, erosion, soil biodiversity, contamination, soil sealing and soil salinisation. Soil organic matter forms the carbon store which is the fundamental element for living fauna in soils that also controls other chemical and physical processes. Intensive agriculture with heavy machinery and large livestock units has also caused soil compaction which is related to factors such as soil texture, packing density, moisture content and plasticity. Compacted soils are also more prone to erosion due to greater ease of runoff associated with the lack of surface roughness and a reduction in tendency for water to seep through the soil profile. Soil biodiversity acts as the engine that makes soil alive and functional in the many processes that it governs. Soil biodiversity is influenced by a wide range of factors including the physico-chemical environment (*e.g.* pH, temperature, water); the supply and availability of soil plant-derived inputs; managed inputs and indigenous organic matter; and the soil habitat. The sources of soil contaminants can be from diffuse or point source pollution and can be broadly classified as nutrients, heavy metals or organic pollutants. These soil contaminants can influence soil biodiversity. Soil sealing and salinisation only affect a small component of agricultural land.

*Corresponding author

Issues in Environmental Science and Technology, 34
Environmental Impacts of Modern Agriculture
Edited by R.E. Hester and R.M. Harrison
© The Royal Society of Chemistry 2012
Published by the Royal Society of Chemistry, www.rsc.org

1 Introduction to Soil Quality

Globally arable and grassland cover areas nearly 16 and 30 million km^2, equivalent to the size of South America and Africa, respectively. Nearly 40% of the surface area of the earth is devoted to agriculture which is more than urban and suburban areas.[1] The agricultural activity results in 30% of global greenhouse gas emissions and 70% of global water withdrawal.[1] Consequently, the impact of agriculture is significant and this pressure is going to increase with the growth in global population and demand for food production. Soils are literally the foundation of terrestrial ecosystems, since they provide the physical surface on which such systems are founded, and support a myriad of functions which deliver ecosystem goods and services. Soils are hence irrevocably fundamental to agricultural production and soil quality is therefore one of the key components that underpins viable and sustainable agricultural production.

Soils are essentially porous media comprised of a wide variety of solid, semi-solid, liquid and gaseous constituents[2] and are arguably the most complex systems on Earth.[3] The solid phases of soils are classed as inorganic ('mineral') or organic (containing carbon). The organic phase is made up of non-living or living matter (the 'biota'). Soils show remarkable spatial heterogeneity across many orders of magnitude of scale, and encompass biological, chemical and physical dynamics over timescales which range from seconds to millennia. Soils are formed by gradual processes of 'weathering' of parent (geological) material, a process which involves a range of chemical, physical and biological mechanisms that result in variably-sized populations of mineral particles being produced. The small, medium (micrometre) and large (millimetre) components are classified as the 'clay', 'silt' and 'sand' fractions, collectively termed 'textural classes'. Soil texture is then a description of the soil in relation to the relative proportions of these size fractions. Sands, silts and clays typically possess a range of contrasting properties in relation to mineral structure and associated physico-chemical attributes, particularly in relation to surface properties and electrostatic charge. These mineral constituents combine with organic components, predominantly originating from green plants, to form relatively thin topsoils which overlay deeper strata or subsoil zones, characterised by lower concentrations of organic material and generally greater bulk density than their upper counterparts. The basic constituents of soil are arranged in space to form a porous network of remarkable properties. The origins of soil pore networks are that the fundamental sand–silt–clay components, mediated by the panoply of organic materials, aggregate to form larger units.[4] These units then aggregate further to create larger structures, with an associated hierarchy of scale and structural stability. Since the aggregates are not uniform in shape, their packing creates a porous matrix, and since the constituents vary widely in size, the resultant porous network is heterogeneous across a concomitantly wide range of scales, exceeding that which occurs in most other porous media. This exceptional heterogeneity of the soil pore network imparts some very significant properties to the soil system. The connectivity and tortuosity of the network governs the movement of gases, liquids and associated solutes, as well as particulates and organisms, through the matrix.[5] This also provides particular mechanisms for the soil matrix to retain water under gravity. Capillary forces which operate in small pores mean that water is retained therein despite gravitational pull, or suction from plant roots, inversely proportionate to the size of the pore. Consequently, the availability of such water varies and in turn modulates processes associated with hydration, such as the dissolution and transport of solutes, biological activity and water availability to plants.[6]

Hence, soils serve several functions which include: productivity or the capability to produce food and fiber; storage, filtration and transformation of materials; a habitat and gene pool of living organisms; a physical and cultural environment for humankind; a source of raw materials (such as peat), acting as a carbon pool; as well as an archive of geological, climatic and archaeological heritage. These functions can be affected by certain threats associated with agricultural activities which are summarised in Figure 1.[7] The following sections of this chapter further unravel these threats in a detailed manner.

According to a recent study report,[1] the current state of agricultural practice is not sustainable for both the production of crops and the environment. This can be associated with the threats which affect soil functions. The suggestion proposed by this team of researchers is an approach known as 'terraculture', or farming for the whole planet which aims to bring the best principles of Green Revolution type commercial agriculture, organic farming and conservation agriculture together to promote a collective and sustainable approach to agriculture.

The importance of protecting soil resources is increasingly recognised. For example, the European Union has adopted the Thematic Strategy for Soil Protection[8] and is considering a draft Soil Framework Directive.

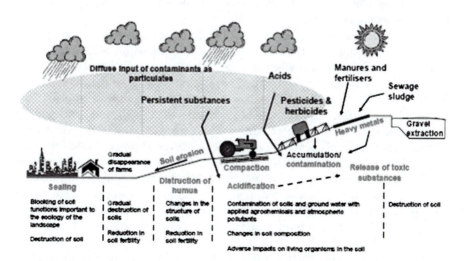

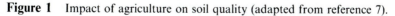

Figure 1   Impact of agriculture on soil quality (adapted from reference 7).

2   Soil Organic Matter Decline

Soil organic matter (SOM) is the product of microbial decomposition of plant tissues. It is highly heterogeneous, reflecting the wide range of materials from which it forms and the complexity of the soil ecology that produces it. In addition to carbon, hydrogen and oxygen, SOM has nitrogen, sulphur and phosphorus in its many structures, with carbon and nitrogen generally in a ratio of 10 to 20. SOM represents the largest stock of terrestrial carbon, estimated[9] at about 2500 Gt or approximately 4.5 times the biotic pool; the annual flux of carbon to and from soils is in the region of 0.5 to 1.0 Gt of carbon.[9] Therefore, the management of SOC, including that in agricultural systems, has profound implications for global carbon trajectories.

The most important role of SOM is as a substrate for soil microbes. The latter support a range of processes essential for the overall functioning of the soil system, including the decomposition of plant materials to release nutrients and allow their recycling, and the physical maintenance of soil aggregates. Plant litter and roots are quickly colonised by microbes to produce products that are in turn further oxidised by other microbes and so on, until the final product is inert, in the sense that the net energy released by further microbial oxidation of the remaining chemical structures is unfavourable. Thus SOM can be viewed as a dynamic continuum of organic molecules, with skeletal structures including rapid turnover polysaccharides, relatively inert materials including polyphenols and inert polymethylene-like ones.

The SOM content of soil varies widely from peat soils (*e.g.* Histosols) where it is the dominant component, to dry sandy ones (*e.g.* Arenosols) where the SOM content is less than fractions of a percent by weight. The level of SOM in soil reflects a dynamic equilibrium between the rates of input from vegetation and oxidation by microbes. The major factors controlling these rates are soil moisture and temperature, the clay content of the soil, and especially the land cover and its management.

Microbial activity is suppressed in cold and wet soils compared to warm and dry ones. Therefore, climate has an important influence on SOM levels. Soil wetness is critical, because when the soil pores are filled with water the soil conditions are less aerobic, and sometimes even anaerobic, so that oxidation of SOM is slowed. This also influences soil temperature, especially below the immediate soil surface that is exposed to solar radiation. As the soil wets up in the autumn, the pores fill until water starts draining from the base of the soil profile. The soil is then at 'field capacity' and remains in this condition until the following spring. Only when the soil starts to dry out and pores are once again filled with air does the soil temperature rise and the SOM decomposition accelerate. Thus SOM levels are better maintained in soils in regions with wetter autumns and springs. In continental and boreal climates, where the soil is frozen to some depth in winter, SOM decomposition is halted for much of the year. These soils tend to have high levels of SOM, often to a considerable depth (*e.g.* Chernozems).

Absorption of SOM to the surfaces of clay minerals protects them from microbial decomposition. Therefore, soils with higher clay contents tend also to have higher SOM levels. This is more likely where the clay minerals are expansive, and shrink and swell as the soil dries out in the spring and wets in the autumn. This causes macropores and fissures to open in the soil profile, allowing root penetration to depth and inputs of plant material throughout the profile. In tropical climates, the soil profile is constantly mixed in soils (*e.g.* Vertisols) with a high concentration of expansive clay minerals ensuring that organic material and SOM are evenly distributed.

Land cover that provides a high Net Primary Production (NPP) from photosynthesis will tend to increase SOM levels. As a general rule, the order of NPP is forest > grassland > arable land. Consequently, where forest is cleared for agriculture a large loss of SOM is normal. Similarly, when permanent grassland is converted to arable production, the inputs of plant materials tend to reduce and SOM levels fall. Tillage to prepare seedbeds, remove weeds and deal with compaction aerates the soil and increases microbial oxidation of SOM, which leads to further losses of SOM. This progression of land use change leading to declining SOM has occurred since the beginning of agriculture and continues today. While there are many field-scale observations of this decline, reliable quantitative estimates of change over large areas (regional to continental) are difficult to obtain and not many are available due to the high temporal and spatial variability in the rates of SOM change.[10] However, several studies[11] indicate that even in apparently mature agricultural landscapes (such as European ones) there are continuing losses of SOM in some arable and grassland systems.

From the above discussion, the complexity of factors controlling levels of SOM in agricultural soils is revealed. Natural factors, including climate and clay content, control the range of SOM contents that specific soils exhibit. Land use is the overriding control on inputs of organic matter and the levels of SOM that are maintained in equilibrium with microbial oxidation.

Once SOM levels have fallen below a critical level, agricultural production starts to fail unless there is substitution of nutrients stored in SOM by fertilisers and inputs such as mechanical tillage and pesticides. This substitution is possible and efficient within modern agriculture, albeit mainly derived from fossil fuel usage, but the loss of SOM is critical in subsistence agricultural systems, particularly where population density and land availability precludes the conversion of further land to cropping. Turning to the management of SOM within developed agricultural systems, there is a widespread view that maintaining SOM levels is beneficial to yields and longer-term productivity. However, the evidence for this is surprisingly limited for conventional agriculture employing fertiliser and other chemical inputs and modern machinery.[12] A substantial review of potential benefits of SOM did not identify significant benefits or provide definitive evidence of a level of SOM below which conventional agriculture might be seriously compromised. It is possible that the main benefits of SOM in conventional agriculture occur infrequently over some years and that they are difficult to discern over the normal periods of experimentation which is only a few years. However, it is clear that maintaining SOM levels in organic farming systems is critical, mainly because SOM provides a continuing reservoir of nitrogen that is essential for sustaining yields. Methods for maintaining SOM levels in arable systems include: introducing grass leys in to crop rotations; planting and incorporating cover crops; reducing the removal and especially the burning of plant residues following harvest; adopting minimum or zero-tillage systems; optimising nutrient and other inputs to crops to maximise organic matter inputs from root and other crop residues. While SOM levels are generally higher under grass, they may be challenged where the removal of dry matter for forage is intensive (by grazing, mechanical harvesting, or both) and where there is regular re-establishment by tillage and re-seeding.

The vast store of carbon in global stocks of SOM means that small changes in its level impact strongly on CO_2 concentrations in the atmosphere. At present, global soils are a net sink for carbon, although they are predicted to become a net source in the second-half of this century.[13] Historic losses of SOC from agricultural soils have been estimated at between 42 and 78 Gt.[9] There is a large potential for carbon sequestration in agricultural soils where these are converted to grasslands or woodlands, and to some extent following adoption of less intensive arable production. In addition, biochar addition to soil has been promoted as a means of sequestering atmospheric carbon.[14] However, there are considerable difficulties in measuring changes in SOC over substantive areas for periods less than about five years, due to the spatial variability in rates of change and the superposition of annual cycles in SOM over longer term trends[10] indicating that opportunities are limited for truly verifiable sequestration within carbon trading and other legal frameworks.

3 Soil Compaction

Soil compaction is the term used to describe a physical reduction in pore space expressed as a reduction in soil volume. Compaction directly affects the complex network of pore spaces and interconnected voids and channels formed by skeletal material consisting of mineral and organic matter, technically referred to as 'soil structure'. The soil structure forms the conduits along which water and gasses flow and exchange, it also provides space through which roots can grow and a habitat for soil organisms.

The susceptibility of a soil to be compacted depends both on the soils physical resilience against compaction, which is related to factors such as soil texture, packing density, moisture content and plasticity, and the forces or pressures that the soil is exposed to that may cause compaction. This relates to factors such as weight of vehicle, distribution of weight on the surface and number of animals in a field. Figure 2 shows an example of soil compaction caused by vehicle wheels. Natural processes can also compact soil, for example natural settlement over time or overburden from the soil above.

Compaction can occur in all soil types, however, in some soils it is more obvious than others. Soils with a moderate to high clay and organic matter content tend to have a good soil structure consisting of a range of pore sizes from micropores (which hold on to water) through to macropores (pores visible without magnification which allow rapid drainage of water and movement of air).

Figure 2 Soil compaction caused by vehicle wheels.

Figure 3 Soil structure in a clay soil (left hand image shows good soil structure and right hand image shows degraded soil structure).

In these soils, compaction which causes the preferential collapse of macro-pores is clearly defined by an absence of these macropores (Figure 3). The soil in the compacted zone is harder to break and breaks into larger pieces, and will offer more resistance to penetration than uncompacted soil. In comparison, compaction in sandy soils is much less obvious as the structure is generally weaker and therefore the range of pore size spaces is less variable. Sand particles because of their size, pack together leaving an open freely draining soil structure.

It has been suggested[15] that compaction in sandy soil manifests itself as tightly interlocking sand particles forming a ridged impenetrable structure. While the change in soil structure is less visually apparent, compacted layers in sandy soils will offer more resistance to penetration and its presence may be apparent by a thickening of root ends near the zone of compaction.

Some soil compaction can bring benefits such as faster seed germination and better water retention that facilitate plant growth. Compaction before tillage can result in a greater surface roughness of the tilled soil that increases infiltration and reduces runoff. However, over-compaction of soil affects productivity and soil function. Over-compaction reduces the rate of water infiltration and drainage, which reduces water storage capacity and increases the likelihood of overland flow following rainfall, increasing the risk of flooding and soil erosion. Changes in soil water storage and gas movement due to over-compaction can affect nutrient cycles and increase greenhouse gas emissions. Soil compaction also increases soil resistance to being moved, making it harder for roots to penetrate (*i.e.* it restricts the roots ability to forage for nutrients and water) and requiring more power (energy) to till the soil.

There are three ways of dealing with compaction: it can be prevented, the soil can be better protected or it can be alleviated. By improving equipment design and soil management practices, soil compaction can be prevented by reducing the pressure exerted on the soil. This can be achieved by: distributing the weight of a vehicle over a wider tyre width, reducing tyre pressure, reducing the number of times the field is driven across, reducing stocking density, accessing the land using tramways, improving the timing of operations to prevent accessing land when the soil is wet and removing animals from fields when the soil is wet. In addition, crop design may be used to reduce the risk of compaction, for example, by changing to crop varieties that can be sown later and/or harvested earlier in the year. However, in extreme cases a change in land use may be the only way to prevent compaction.

To protect the soil from compaction the soil needs to be made more resilient. Soil resilience can be improved by increasing the soil organic matter content, improving drainage and minimising tillage operations (*i.e.* using conservative tillage techniques that disturb the soil structure less). The soil can also be protected by reinforcing the surface with geotextile materials (made from natural materials: coirs, jute, hemp, straw, or synthetic materials: polypropylene) at focal points such as feeding and drinking troughs or along trackways.

If soil compaction does occur, it will need to be alleviated. Some natural recovery from shallow (>30 cm depth) soil compaction may occur depending on: the severity of compaction, duration of time the soil has to recover, soil type and climatic conditions (*i.e.* freeze–thaw or shrink–swell cycles). More frequently, mechanical intervention will be required to alleviate soil compaction. Ploughing an arable field will remove compaction from topsoil (>30 cm depth), however, deeper subsoiling will be needed to remove compaction from subsoil layers (30 to 60 cm depth). In grassland deep subsoiling will destroy the sward, therefore shallow grassland subsoilers and sliters (aerators) are used to varying degrees of success.[16,17]

4 Soil Erosion

Soil erosion is the removal of soil from a location resulting in a reduction of soil mass and soil volume. The process of erosion involves stages of soil detachment, entrainment, transport and deposition, in a cycle that is repeated until the erosive energy is depleted. Soil erosion can occur under all agricultural land uses but is usually more apparent in arable fields. In the agricultural landscape, four forms of erosion can be defined: soil erosion by water, wind, tillage and co-extraction with root crops and farm machinery. Figure 4 shows an example of erosion caused by water resulting in an erosion feature known as a 'rill'.

Soil erosion by water and wind are the most widely studied. In many respects, they are very similar processes, both demonstrate an increase in erosivity as water and wind speed increase. Soil texture, organic matter content, surface roughness and land cover all affect the amount of soil erosion by water and wind. However, the processes differ in the extent of area that is affected. Erosion by wind may affect the whole field, while soil erosion by water primarily occurs at focal points in the field. Although, soil erosion by water can occur over a larger area as 'sheet erosion', the roughness of the soil surface limits detachment and transport. Rate of erosion by water is also affected by slope steepness, with greater erosion occurring on steeper slopes. Erosion by water and wind can both remove soil from a field and cause wider environmental damage.

Tillage erosion can move one thousand times more material downslope than natural soil creep.[18] The rate of soil movement depends on slope gradient, slope curvature, plough depth, plough direction, ploughing speed and plough design. The greatest soil movement occurs on steep convex slopes when the soil is ploughed in the downslope direction. However, unless the downslope field boundary is removed, there is no export of soil from the field.

Figure 4 An example of soil erosion by water.

Very little is known about the co-extraction of soil with root crops and farm machinery.[18] Soil is removed from the field on root crops such as sugar beet, potatoes, carrots and chicory root, and on the wheels and implements (*e.g.* tines, ploughs and discs) of farm machinery. The amount of soil removed depends on a range of factors, including: soil texture (clay being most cohesive), organic matter, calcium carbonate ($CaCO_3$) content, soil moisture and soil structure. Crop factors and harvesting techniques also affect the amount of soil removed, including: roughness of skin, shape of tuber, type of machines used to lift the crop and in-field processing to remove soil from the crop.

Soil erosion caused by any of these four processes affects soil function in a number of ways. Primarily the erosion removes valuable topsoil that contains nutrients. The removal of these nutrients can be compensated for by adding fertiliser but is not a sustainable solution. Also, other soil functions are affected by the removal of soil. Loss of topsoil also reduces media for root growth and exploration, soil water storage capacity, organic matter, soil habitat, buffering capacity, and the capability for the filtration and transformation of chemicals and organic wastes, and can expose archaeological remains making them vulnerable to deterioration. Soil erosion can also have a wider environmental impact, such as eutrophication of rivers and lakes by nutrients carried with the sediment.

In-field reduction of soil erosion by water and wind can be achieved by maintaining good surface cover and surface roughness especially at vulnerable times, for example establishing a cover crop over winter or by leaving stubble on the field after harvest. By maintaining a high organic matter content and encouraging good soil structure, the resilience of the soil to being eroded can also be improved. Changes in land management practices, for example, from conventional to conservative tillage, have been shown to reduce soil loss because of better surface protection and a more stable soil structure.[19] In some cases, especially in relation to water erosion at a focused point, it may be necessary to provide additional surface protection using geotextiles to initially stabilise the soil before a cover crop, such a grass, can be established to help reduce soil loss.

To reduce the amount of soil erosion by tillage, the amount of downslope ploughing needs to be minimised and ploughs designed to minimise downslope movement utilised. While soil loss through co-extraction can be reduced by harvesting crops when the soil is dry, it is desirable to have better on-site processing of the crop to remove more soil before it leaves the field and to use new varieties of crop that trap less soil on their surface.

5 Soil Biodiversity

An immense quantity and range of life resides in the upper layers of most soils. The statistics are now well rehearsed but always remarkable. For example, the total fresh weight mass of the biota below old temperate grassland can exceed 45 t ha^{-1}.[20] A few grams of such soil contains billions of bacteria, hundreds of kilometres of fungal hyphae, tens of thousands of protozoa, thousands of

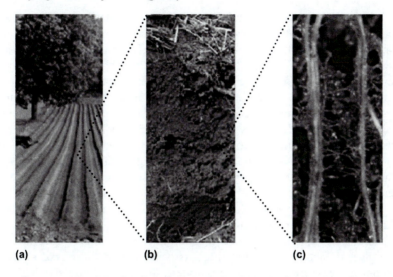

Figure 5   Magnification of various scales ranging from (a) field to (b) soil surface and (c) roots and bridging fungal hyphae indicating soil biota as the foundation of soil systems.

nematodes, several hundred insects, arachnids and worms, and hundreds of metres of plant roots. These large numbers are matched by extreme levels of biodiversity, most particularly at the microbial scale. Figure 5 shows an illustration of the significance of microbial scale in relation to a larger field scale. Whilst the biomass typically only constitutes a small proportion of the total mass of soils, it has a hugely disproportionate effect upon soil functions. The entirety of the soil biomass, whilst predominantly microbial, interacts *via* a series of complex food webs involving energy transfers as organisms within a range of trophic groups consume and are consumed by other organisms.[21] Microbes function as primary decomposers and biochemical transformers at the core of such webs, and larger organisms provide higher-order ecosystem services, such as organic matter comminution, decomposition and ecosystem engineering. The soil biota can thus be conceived of as the 'biological engine of the earth', driving many of the key processes that occur within soils. Some specific aspects of biotic contributions to soil function are described throughout this chapter. This section then focuses primarily on the effects of agriculture on biodiversity *per se*, rather than such processes.

The biodiversity of communities can be defined as both the basic number of taxonomically distinct organisms ('richness'), or as a more literal constitutional composition, which is sometimes referred to as 'community structure'. The functional consequences of species richness in soils are not well understood. Biodiversity is considered to provide an 'insurance' against environmental change.[22] It is generally hypothesised that greater richness leads to greater resistance to effects of environmental perturbation upon system function, and a concomitantly enhanced capacity for recovery following such disturbance.[23]

This is generally the case, however, experimental evidence in this respect is equivocal, and it is increasingly apparent that community structure is more pertinent in this respect, *i.e.* the identity, and specifically the functional traits, of the organisms present, rather than the richness itself.[24] This means that it is difficult to generalise about the effects of changes in biodiversity (richness or structure). In general, intensive agricultural systems which rely upon industrially-based substitutions for ecological services and which are not reliant on biologically-mediated functions result in a decline in richness across the biotic spectrum, simpler community structures, and less complex trophic webs.[25] Such effects may not influence productivity (due to the substitution principles) but can have implications for longer term sustainability.[26] Conversely, production systems which rely upon such biological bases to function support greater richness and more diverse trophic webs, and the community structures therein are important for system function.

Soil biodiversity, and particularly community structure, is governed by a wide range of factors including: those associated with the physico-chemical environment (*e.g.* pH, temperature, water); the supply and availability of energy to the biota (governed principally by soil plant-derived inputs, managed inputs and indigenous organic matter); and the soil habitat, manifest principally as the pore network discussed above. Hence any agricultural practices which affect such factors will duly influence biodiversity, and can therefore have both direct and indirect effects upon soil biota. In general, heterogeneity (physical, chemical, climatological in both space and time) begets biodiversity, and thus the more monotonous the production system, the lower the inherent diversity.

Direct effects notably include those induced by the application of biocides which target specific biotic groups which are deemed to impair crop production, including: herbicides, pesticides, nematicides and fungicides.[26] As well as affecting such biota directly, there are often effects on non-target organisms both within the same and other biotic groups. Contemporary and emerging legislation in Europe is leading to the withdrawal of potent broad-spectrum biocides, and a trend for increased specificity in such agents, as well as lower effective concentrations, will likely reduce such effects.

Other examples of direct effects include: modification of soil pH *via* liming, which by raising pH will favour community structures adapted to less acid conditions, and application of inorganic fertilisers which will tend to suppress free-living N-fixing organisms. Tillage practices which disrupt the soil matrix, such as ploughing and harrowing, will directly affect organisms which have a high reliance on structurally intact soils, such as surface-casting earthworms, since their burrow networks are disrupted.[27] Tillage also tends to reduce the relative proportion of fungi in soils since the filamentous networks of fungal hyphae (mycelia) (as shown in Figure 5) are also disrupted.[28] In the case of both burrows and mycelia, regeneration in the short term can occur, but persistent disruption compromises these organisms and their associated diversity.

Indirect effects of agricultural practices upon soil biodiversity are arguably of greater influence since they are manifest at a greater scale and operate at a broader system-level. Examples here include the nature of the crop grown, both in any one season or in the context of any rotational practice,[29] application of organic materials that serve as energy sources and/or mineral nutrients, and tillage practices insofar as they affect the architecture of the soil.[30]

6 Soil Contamination

There are various sources of contaminants which include diffuse and point-source pollution. They include discharges from farming activities (*e.g.* sewage sludge and manure application, sheep dipping chemicals *etc.*) and atmospheric deposition. In addition to these sources of contaminants which are more conventional, there seems to be increasing concern on emerging pollutants. Table 1 summarises the various sources of contaminants in the environment which will be discussed in detail below.

Diffuse soil contamination results from no one discrete source but is due to widespread activities. Agricultural soil can be contaminated by atmospheric deposition and other chemical particles. It can also be contaminated by deposition of contaminated sediments from flood water and by agricultural practices such as the application of fertilisers, manure, pesticides and biowaste

Table 1 Sources of contaminants into the aquatic environment (adapted from references 31 and 40).

Emission type	Contaminant class	Major examples
Diffuse – agriculture and urban	Pesticides	Herbicides (*e.g.* glyphosate, atrazine, simazine, isoproturon), fungicides
	Nutrients	Nitrates, phosphates
	Heavy metals	Cadmium, copper, zinc, lead and arsenic
	Human pharmaceuticals	Carbamazepine, chorhexidine, fluoxetine, cotinine, fluoxetine
	Veterinary medicines	Antibiotics, coccidiostats, ecto- and endo- parasiticides
	Personal care products	Methyl paraben, sodium lauryl sulfate, triclosan
	Persistent Organic Pollutants	Dioxins, furans, PCBs, aldrin, chlordane, DDT
Point-source – agriculture	Pesticides	Herbicides (*e.g.* glyphosate, atrazine, simazine, isoproturon), fungicides
	Veterinary medicines	Growth hormones (manure borne estrogens), roxarsone (arsenic-containing feed additive in poultry), coccidiostats (to overcome poultry intestinal parasites), antibiotics

to the land. The main forms of agricultural soil contamination are therefore contaminations with nutrients, heavy metals and organic pollutants.

6.1   Nutrients

Nitrogen and phosphorus are particularly important as both are implicated in aquatic eutrophication. Eutrophication and the associated ecological effects result in general decline in overall water quality, restricting its use for general and drinking purposes. Nutrients in agricultural runoff can arise from point or diffuse sources of pollution, with major point-sourced pollution incidents occurring due to poor containment of slurry or silage effluents. Such point sources of pollution are easy to identify and control. However the diffuse sources of pollution, such as losses of nutrients through leaching and in surface runoff (due to slurry, manure or fertiliser application on fields or unsafe disposals of sheep dips) are more difficult to assess and control.[31]

Surplus nutrients in the soil can accumulate when excess nutrients are added beyond what is needed by the crop. This can be due to overestimating crop requirements or to undervaluing the nutrient component of organic amendments that may be added to a field in addition to inorganic fertiliser simply as a means of disposal. When organic amendments are applied to soils, the availability of nutrients for crop uptake is slower than inorganic fertilisers. This is because the nutrients in organic amendments need to be made available by soil microbial communities which occurs along a longer pathway. Consequently over a period of time soils that have received repeated application of organic amendments can be enriched and any further additions of inorganic fertilisers need to take into account historical applications of organic amendments. The build-up of such nutrients has little effect on productivity but are vulnerable to release to the wider environment where their presence may cause biological damage *e.g.* eutrophication of rivers and lakes.

Diffuse soil contamination decreases soil quality affecting soil functions relating to provision, filtration, cultural and support. Soil contamination is of particular concern because of its relationship with food production and therefore the potential of contaminants to enter the food chain. Contamination has the potential to reach the human population directly through the food they eat. Crops may take-up harmful substances through their root systems while animals may ingest contaminated fodder crops or supplements. Soil contamination can also affect other food chains through the consumption of contaminated soil organisms that are eaten by higher order animals. The effects of contamination on the soils ability to filter, buffer and transform chemicals is also of concern because of the wider environmental significance this has, including the release of contaminants into groundwater and surface water supplies.

Diffuse soil contamination may be caused by a slow accumulation of chemicals over time or may be due to a single identifiable event such as accidental spillage or flood deposition. Prevention is easier than dealing with the consequences, therefore, regulation that reduces potential pollution delivery by airborne and flood deposition, better agricultural management to ensure optimal nutrient management and improved monitoring schemes to detect potential problems at the earliest opportunity will all help to reduce contamination risk.

6.2 Heavy Metals

Soil acts as a long term sink of heavy metal accumulation as a net result of the different mobility and bioavailability of heavy metal in soils, leaching losses and plant uptake. The significance of the accumulation of heavy metals in soils can impact water supply *via* borehole abstraction and associated groundwater activities.

Heavy metal input to agricultural soils originates from various sources including atmospheric deposition, biosolids, livestock manures, inorganic fertilisers and lime, industrial by-products and composts. These sources of heavy metal input can contribute to significant levels of zinc, copper, nickel, lead, cadmium, chromium, arsenic and mercury.[32] Atmospheric deposition was the main source of most metals entering agricultural land representing 49% of total zinc inputs of 5040 t year^{-1}, 39% of copper inputs of 1620 t year^{-1} and 54% of cadmium inputs of 40 t year^{-1}. Livestock manures and biosolids were also important sources representing 37 and 8% of total zinc inputs, 40 and 17% of copper and 10 and 4% of cadmium inputs, respectively.[32]

The heavy metals get in to excreta *via* food supplements fed to the animals *e.g.* copper and zinc in pig manure is a consequence of the mineral supplement in the pig diet. Accumulation of heavy metals in the soil can harm soil flora and fauna leading to changes in soil community structure that impact on soil formation through changes in organic matter decomposition. It can also disrupt nutrient cycles because of loss of organisms involved in the cycling process.

6.3 Organic Pollutants

Pesticides in soils are widely studied because they are commonly used to control pests that affect agricultural crops and pests in the home, yards and gardens. The fate of pesticides in soil is controlled by chemical, biological and physical dynamics of this matrix. These processes can be grouped into those that affect persistence, including chemical and microbial degradation and those that affect mobility, involving sorption, plant uptake, volatilization, wind erosion, runoff and leaching.

Pesticides are degraded by chemical and microbiological processes. Chemical degradation occurs through reactions such as photolysis, hydrolysis, oxidation and reduction. Biological degradation takes place when soil microorganisms consume or break down pesticides. These microorganisms are mainly distributed in the top centimetres of the surface layer of the soil, where the organic matter acts as food supply. The extent of degradation ranges from formation of transformation products (TPs) to decomposition in inorganic products.[33] TPs can be present at higher levels in the environment than the parent and in some instances can be more persistent, more mobile and more toxic.[33,34]

In addition, some pesticides are chiral molecules. In that way, environmental studies have historically neglected to determine the adverse effects associated with particular enantiomers, including persistence in various environmental media. The racemic signature remains unchanged by physico-chemical removal mechanisms. However, microbial degradation and biological metabolism may be enantio-selective, and result in different effects and fates in the environment.[35]

The movement of pesticides is determined by solubility, soil-sorption constant (K_{oc}), the octanol–water partition coefficient (K_{ow}) and half life in soil (DT_{50} or $T_{\frac{1}{2}}$). Pesticides can enter aquatic systems *via* spray drift surface runoff or drainflow. The potential for transport to aquatic systems is highly dependent on the chemical properties and the nature of the soil environment. A pesticide is able to contaminate groundwater (leaching) if its sorption coefficient is low, its half-life long and its water solubility high. This is quite frequent because pesticides are increasingly polar, hydrosoluble and thermolabile to diminish their toxicity and to facilitate their disappearance from the environment; at the same time, they must persist long enough to enable acceptable pest control.

Veterinary medicines may also accumulate in the soil through the application of manure to the land.[36] Veterinary medicines may also affect soil organism community structures and thus have similar effects on soil function as heavy metals do. Veterinary medicines are widely used to treat disease and protect the health of animals. Dietary enhancing feed additives (growth promoters) are also incorporated into the feed of animals reared for food in order to improve their growth rates. Release of veterinary medicines to the environment occurs both directly, for example the use of medicines in fish farms, and indirectly, *via* the application of animal manure (containing excreted products) to land or *via* direct excretion of residues onto pasture.[34,37,38]

Organic pollutants that are emerging can be associated with chemicals that are associated with personal care products. Table 1 summarises some of the products. In general most of these products end up in the soil as part of added biosolids to land. The removal of emerging organic pollutants is challenging using conventional treatment processes in the sewage treatment plants. Consequently these organic pollutants are accumulated in biosolids and end up in the terrestrial ecosystem when applied to arable soils. A recent study[39] reports that triclosan (which is an antibacterial agent in tooth paste, mouth wash, after shave and shampoos) when applied to arable soils as part of biosolids influences soil microbial respiration. However the resilience of the soil microbial community recovers after a period of time and the added biosolids which has some triclosan does not completely diminish the soil ecology.

7 Soil Sealing

Soil sealing occurs when agricultural land is taken in to the built environment. It is a major threat to the sustainability of agriculture because it is essentially irreversible. The built environment is expanding rapidly, driven by growth in the urban population which is forecast to increase from 3.49 billion in 2010 to 5.26 billion by 2030 (United Nations, 2010). Within Europe, the area of land covered by artificial surfaces (*e.g.* for residential areas, industrial and commercial sites) increased by 6258 km^2 (3.4%) from 2000–2006.[41] Soil sealing occurs on some of the most valuable soils for agriculture since urban centres were often founded near rivers and estuaries and close to fertile land that could provide sufficient food supplies. Today, appropriate development controls are needed to protect valuable agricultural soils and ensure food security for the urban population.

8 Soil Salinisation

Salinisation is the excessive concentration of water soluble salts in soil so that it no longer supports healthy plant growth. If the salts are predominantly sodium salts, the term 'sodification' is used. Salts accumulate in the soil profile when evapotranspiration exceeds precipitation. Therefore, its natural occurrence (as Solonetz and Solonchak soils) is limited to certain climatic regions, namely arid, semi-arid and some continental and Mediterranean zones. Extensive areas of soil in these zones are salt-affected due to inappropriate agricultural management. Typically, a threshold for excess salt content would be 0.1% total salt content or a conductivity of $4\,dS\,m^{-1}$ in the top 0.3 m of the soil profile.[42] The salt composition is important with higher exchangeable sodium content (as a percentage of total salts) leading to greater crop yield losses because of excessive alkalinity and structural degradation caused by the dispersal of clay particles. The main issue is the use of irrigation water that contains soluble salts and/or its poor management, including problems arising from rising saline groundwater. Irrigation water with a conductivity of greater than $0.5\,dS\,m^{-1}$ needs to be used carefully, taking account of the crop type and its salt tolerance. Salt-affected soils sometimes can be recovered for agriculture by improved irrigation management (where there is a suitable supply of water) and by addition of gypsum ($CaSO_4$) but often this is impractical.

9 Conclusions

Agriculture covers nearly 40% of the surface area of the earth and results in 30% of global greenhouse gas emissions and 70% of global water withdrawal. Consequently the impact of agriculture is significant and this pressure is going to increase with the increment in global population and demand for food production. Soils are hence irrevocably fundamental to agricultural production and soil quality is therefore one of the key components that underpins viable and sustainable agricultural production. There are certain threats to soil quality associated with agricultural activities which are decline in soil organic matter, compaction, erosion, soil biodiversity, contamination, soil sealing and soil salinisation. The key approaches to minimise each of these threats can be summarised as follows:

i. Adequate level of SOM is essential to ensure viable agricultural production. Maintaining SOM levels is critical, mainly because SOM provides a continuing reservoir of nitrogen that is essential for sustaining yields. Methods for maintaining SOM levels in arable systems include: introducing grass leys in to crop rotations; planting and incorporating cover crops; reducing the removal and especially the burning of plant residues following harvest; adopting minimum or zero-tillage systems; optimising nutrient and other inputs to crops to maximise organic matter inputs from root and other crop residues. The vast store of carbon in global stocks of SOM means that small changes in its level impact strongly on CO_2 concentrations in the atmosphere.

ii. There are three ways of dealing with compaction, it can be prevented, the soil can be better protected or it can be alleviated. By improving equipment design (*e.g.* tyre pressure) and soil management practices (*e.g.* reducing livestock density) soil compaction can be prevented by reducing the pressure exerted on the soil. To protect the soil from compaction the soil needs to be made more resilient. Soil resilience can be improved by increasing the soil organic matter content, improving drainage and minimising tillage operations *i.e.* using conservative tillage techniques that disturb the soil structure less. The soil can also be protected by reinforcing the surface with geotextile materials (made from natural: coirs, jute, hemp, straw, and synthetic: polypropylene, materials) at focal points such as feeding and drinking troughs or along trackways. If soil compaction does occur, it will need to be alleviated. Some natural recovery from shallow (>30 cm depth) soil compaction may occur depending on the severity of compaction, duration of time the soil has to recover, soil type and climatic conditions (*i.e.* freeze-thaw or shrink-swell cycles). More frequently mechanical intervention will be required to alleviate soil compaction.

iii. In-field reduction of soil erosion by water and wind can be achieved by maintaining good surface cover and surface roughness especially at vulnerable times, for example establishing a cover crop over winter or by leaving stubble on the field after harvest. By maintaining a high organic matter content and encouraging good soil structure the resilience of the soil to being eroded can also be improved. Changes in land management practices for example from conventional to conservative tillage has been shown to reduce soil loss because of better surface protection and a more stable soil structure. In some cases, especially in relation to water erosion at a focused point, it may be necessary to provide addition surface protection using geotextiles to initially stabilise the soil before a cover crop, such a grass, can be established to help reduce soil loss.

iv. In general, heterogeneity (physical, chemical, climatological in both space and time) begets soil biodiversity, and thus the more monotonous the production system the lower the inherent diversity. Direct effects notably include those induced by the application of biocides which target specific biotic groups which are deemed to impair crop production, including herbicides, pesticides, nematicides and fungicides. As well as affecting such biota directly, there are often effects on non-target organisms both within the same and other biotic groups. Other examples of direct effects include modification of soil pH *via* liming, which by raising pH will favour community structures adapted to less acid conditions, and application of inorganic fertilisers which will tend to suppress free-living nitrogen-fixing organisms. Tillage practices which disrupt the soil matrix, such as ploughing and harrowing, will directly affect organisms which have a high reliance on structurally intact soils, such as surface-casting earthworms, since their burrow networks are disrupted. Tillage also tends to reduce the relative proportion of fungi in soils since the filamentous networks of fungal hyphae (mycelia) are also disrupted. In the case of both burrows and mycelia, regeneration in the short term can occur, but persistent disruption compromises these organisms and their associated diversity.

v. Soil contaminants can originate from both diffuse and point source pollution and can be broadly classified into: nutrients, heavy metals and organic micropollutants. Soil contaminations associated with nutrients can be managed if farmers apply fertilisers in accordance with crop needs and not in excess. In addition, nutrient supply from organic amendments needs to be considered and reduction made on any added inorganic fertilisers. Heavy metal input to agricultural soils originates from various sources, including: atmospheric deposition, biosolids, livestock manures, inorganic fertilisers and lime, industrial by-products and composts. Organic pollutants in soil originate from added pesticides, herbicides, fungicides, and personal care and pharmaceutical products (originating from added biosolids). Heavy metal and organic pollutants in soil can be reduced by minimising these pollutants at source and also following stipulated guidelines available in the agricultural sector.

vi. Soil threats associated with sealing and salinisation are very specific and only affect a limited area of productive land presently. Soil sealing occurs when agricultural land is taken in to the built environment. It is a major threat to the sustainability of agriculture because it is essentially irreversible. Salinisation is the excessive concentration of water soluble salts in soil so that it no longer supports healthy plant growth.

References

1. J. A. Foley, N. Ramankutty, K. A. Brauman, E. S. Cassidy, J. S. Gerber, M. Johnston, N. D. Mueller, C. O'Connell, D. K. Ray, P. C. West, C. Balzer, E. M. Bennett, S. R. Carpenter, J. Hill, C. Monfreda, S. Polasky, J. Rockström, J. Sheehan, S. Siebert, D. Tilman and D. P. M. Zaks, *Solutions for a Cultivated Plane*t, 2011; doi:10.1038/nature10452 (last accessed 12 October 2011).
2. N. C. Brady and R. R Weil, *The Nature and Properties of Soils,* Prentice Hall, Upper Saddle River, NJ, 2002.
3. K. Ritz, Soil as a paradigm of a complex system, in *Complexity and Security*, ed. J. J. Ramsden, P. J. Kervalishvili, IOS Press, Amsterdam, 2008, pp. 103–119.

4. J. M. Tisdall and J. M. Oades, Organic matter and water-stable aggregates in soils, *J. Soil Sci.*, 1983, **33**, 141–163.
5. I. M. Young and K. Ritz, The habitat of soil microbes, in *Biological Diversity and Function in Soils*, ed. R. D. Bardgett, M. B. Usher and D. W. Hopkins, Cambridge University Press, Cambridge, 2005, pp. 31–43.
6. W. A. Jury and R. Horton, *Soil Physics,* John Wiley, Chichester, 2004.
7. W. E. H. Blum, Characterisation of soil degradation risk: an overview, in *Threats to Soil Quality in Europe*, ed. G. Tóth, L. Montanarella and E. Rusco, JRC Scientific and Technical Report, Italy, 2008 p. 10.
8. European Commission, *Thematic Strategy for Soil Protection, COM(2006)231 Final*, Commission of the European Communities, Brussels, 2006.
9. R. Lal, Soil carbon sequestration impacts on global climate change and food security, *Science*, 2004, **304**, 1623–1627; doi:10.1126/science.109739.
10. N. P. A. Saby, P. H. Bellamy, X. Morvan, D. Arrouays, R. J. Jones, F. G. A. Verheijen, M. G. Kibblewhite, A. Verdoot, J. B. Üveges, A. Freudenschuß and C. Simota, Will European soil monitoring networks be able to detect changes in topsoil organic carbon content?, *Global Change Biol.*, 2008, **14**, 2432–2442; doi:10.1111/j.1365-2486.2008.01658.x.
11. P. H. Bellamy, P. J. Loveland, R. I. Bradley, R. M. Lark and G. J. D. Kirk, Carbon losses from all soils across England and Wales 1978–2003, *Nature*, 2005, **437**, 245–248.
12. P. Loveland and J. Webb, Is there a critical level of organic matter in the agricultural soils of temperate regions: a review, *Soil Tillage Res.*, 2003, **70**, 1–18; doi:10.1016/S0167-1987(02)00139-3.
13. P. M. Cox, R. A. Betts, C. D. Jones, S. A. Spall and I. J. Totterdell, Acceleration of global warming due to carbon-cycle feedbacks in a coupled climate model, *Nature*, 2000, **408**, 184–187.
14. J. Lehmann, J. Gaunt and M. Rondon, *Mitigation Adaptation Strategies Global Change*, 2006, **11**, 403–427.
15. T. Batey, *Soil Husbandry*, Soil and Land Use Consultants, Aberdeen, 1988.
16. J. T. Douglas, C. E. Crawford and D. J. Campbell, Traffic systems and soil aerator effects on grassland for silage production, *J. Agric. Eng. Res.*, 1995, **60**, 261–270.
17. R. A. Fortune, P. D. Forristal and F. Kelly, Effects of soil aeration in minimising/alleviating soil compaction and sward damage in grassland, *Teagasc*, 1999; http://www.teagasc.ie/research/reports/crops/4352/eopr-4352.pdf (last accessed 12 October 2011).
18. P. N. Owens, R. J. Rickson, M. A. Clarke, M. Dresser. L. K. Deeks, R. J. A. Jones, G. A. Woods, K. Van Oost and T. A. Quine, *Review of the Existing Knowledge Base on Magnitude, Extent, Causes and Implications of Soil Loss due to Wind, Tillage and Co-extraction with Root Vegetables in England and Wales, and Recommendations for Research Priorities.* 2006, National Soil Resources Institute (NSRI) Report to DEFRA, Project SP08007, NSRI, Cranfield University, UK.

19. J. M. Holland, The environmental consequences of adopting conservation tillage in Europe: reviewing the evidence, *Agric. Ecosyst. Environ.*, 2004, **103**, 1–25.

20. K. Ritz, Underview: origins and consequences of belowground bio-diversity, in *Biological Diversity and Function in Soils*, ed. R. D. Bardgett, M. B. Usherand D. W. Hopkins, Cambridge University Press, Cambridge, 2005, pp. 381–401.

21. W. H. Van der Putten, P. C. de Ruiter, T. M. Bezemer, J. A. Harvey, M. Wassen and V. Wolters, Trophic interactions in a changing world, *Basic Appl. Ecol.*, 2004, **5**, 487–494.

22. M. B. Postma-Blaauw, R. G. M. de Goede, J. Bloem, J. H. Faber and L. Brussaard, Soil biota community structure and abundance under agricultural intensification and extensification, *Ecology*, 2010, **91**, 460–473.

23. S. Botton, M. van Heusden, J. R. Parsons, H. Smidt and N. H. van Straalen, Resilience of microbial systems towards disturbances, *Crit. Rev. Microbiol.*, 2006, **32**, 101–112.

24. C. Stoate, A. Baldi, P. Beja, N. D. Boatman, I. Herzon, A. van Doorn, G. R. de Snoo, L. Rakosy and C. Ramwell, Ecological impacts of early 21st century agricultural change in Europe - A review, *J. Environ Manage.*, 2009, **91**, 22–46.

25. M. G. Kibblewhite, K. Ritz and M. J. Swift, Soil health in agricultural systems, *Philos. Trans. R. Soc. London, Ser. B*, 2008, **363**, 685–701.

26. R. Ghorbani, S. Wilcockson, A. Koocheki and C. Leifert, Soil management for sustainable crop disease control: a review, *Environ. Chem. Lett.*, 2008, **6**, 149–162.

27. J. Roger-Estrade, C. Anger, M. Bertrand and G. Richard, Tillage and soil ecology: partners for sustainable agriculture, *Soil Tillage Res.*, 2010, **111**, 33–40.

28. K. Ritz and I. M. Young, Interactions between soil structure and fungi, *The Mycologist*, 2004, **18**, 52–59.

29. B. C. Ball, I. Bingham, R. M. Rees, C. A. Watson and A. Litterick, The role of crop rotations in determining soil structure and crop growth conditions, *Can. J. Soil Sci.*, 2005, **85**, 557–577.

30. L. Munkholm and B. D. Kay, Managing the interactions between soil biota and their physical habitat in agroecosystems, in *Architecture and Biology of Soils*, ed. K. Ritz and I. M. Young, CABI, Wallingford, 2010, pp. 170–195.

31. P. S. Hooda, A. C. Edwards, H. A. Anderson and A. Miller, A review of water quality concerns in livestock farming areas, *Sci. Total Environ.*, 2000, **250**, 143–167.

32. F. A. Nicholson, S. R. Smith, B. J. Alloway, C. Carlton-Smith and B. J. Chambers, Quantifying heavy metal inputs to agricultural soils in England and Wales, *Water Environ. J.*, 2006, **20**, 87–95.

33. V. Andreu and Y. Pico, Determination of pesticides and their degradation products in soil: critical review and comparison of methods, *Trends Anal. Chem.*, 2004, **23**(10–11), 772–789.

34. A. B. A. Boxall, D. Kolpin, B. Halling SØrensen and J. Tolls, Are veter-
    inary medicines causing environmental risks?, *Environ. Sci. Technol.*, 2003,
    37(15), 286A–294A.
35. A. Monkiedje, M. Spiteller and K. Bester, Degradation of racemic
    and enantiopure metalaxyl in tropical and temperate soils, *Environ. Sci.
    Technol.*, 2003, **37**(4), 707–712.
36. U. Hammesfahr, R. Bierl and S. Thiele-Bruhn, Combined effects of the
    antibiotic sulfadiazine and liquid manure on the soil microbial-community
    structure and functions, *J. Plant Nutrit. Soil Sci.*, 2011, **174**, 614–623.
37. A. B. A. Boxall, L. A. Fogg, P. Kay, P. A. Blackwell, E. J. Pemberton and
    A. Croxford, Veterinary medicines in the environment, *Rev. Environ.
    Contamination Toxicol.*, 2004, **180**, 1–91.
38. S. E. Jorgensen and B. Halling-Sorensen, Drugs in the environment,
    Chemosphere, 2000, **40**(7), 691–699.
39. E. Butler, M. J. Whelan, K. Ritz, R. Sakrabani and R. van Egmond,
    Effects of triclosan on soil microbial respiration, *Environ. Toxicol. Chem.*,
    2011, **30**(2), 360–366.
40. J. A. Plant, A. Korre, S. Reeder, B. Smith and N. Voulvoulis, Chemicals in
    the environment: implications for global sustainability, *Appl. Earth Sci.
    (Trans. Inst. Min. Metall. B)*, 2005, **114**, B65–B97.
41. European Environment Agency, *The European Environment - State and
    Outlook 2010. Land Use*, Publications Office of the European Union,
    Luxembourg, 2010; doi:10.2800/59306.
42. S. Huber, G. Projop, D. Arrouays, G. Banko, R. J. A. Jones, M. G.
    Kiblewhite, W. Lexer, A. Moller, R. J. Rickson, T. Shishkov, M. Stephens,
    G. Toth, J. J. H. Van Den Akker, G. Varallyay, F. G. A. Verheijen and
    A. R. Jones, *Environmental Assessment of Soil for Monitoring. Volume I.
    Indicators & Criteria*, EUR 23490 EN/1, Office of the Official Publica-
    tions of the European Communities, Luxembourg, 2008, p. 88; doi:
    10.2788/93515.

Impacts of Agriculture upon Greenhouse Gas Budgets

J. M. CLOY,* R. M. REES, K. A. SMITH, K. W. T. GOULDING,
P. SMITH, A. WATERHOUSE AND D. CHADWICK

ABSTRACT

Agriculture and land use change are responsible for one third of global greenhouse gas (GHG) emissions. These emissions have risen steadily during the 20th century and pressures to increase food production to support growing human populations threaten to increase emissions still further, as climate change induced by the accumulation of GHGs threatens the sustainability of agricultural production. There are, however, opportunities to mitigate emissions through modifications to agricultural management that involve more efficient use of inputs and promotion of carbon sequestration. Management activities need to be adapted to take account of the variability in climates and soils, and patterns of emissions. There also need to be improvements to current methods of compiling inventories of GHG emissions to ensure that they accurately reflect actual emissions and management changes.

1 Introduction

Agriculture and land use change are major global sources of greenhouse gas (GHG) emissions. Recent estimates suggest that these activities together contribute nearly a third of global emissions, which makes them second in importance only to energy production (Figure 1).[1] There has been a large

---

*Corresponding author

Issues in Environmental Science and Technology, 34
Environmental Impacts of Modern Agriculture
Edited by R.E. Hester and R.M. Harrison
© The Royal Society of Chemistry 2012
Published by the Royal Society of Chemistry, www.rsc.org

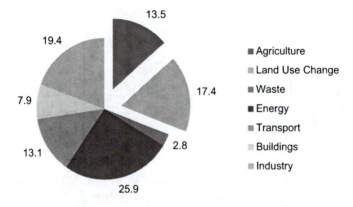

13.5

19.4

7.9

17.4

2.8

13.1

25.9

- Agriculture
- Land Use Change
- Waste
- Energy
- Transport
- Buildings
- Industry

Figure 1   Greenhouse gas (GHG) emissions by source, 2004, expressed as percentages of total carbon dioxide (CO_2) equivalent emissions based on global warming potentials. Overall, agriculture (cropping and livestock) contributes 13.5% of global GHG emissions mostly through emissions of methane (CH_4) and nitrous oxide (N_2O) (about 47% and 58% of total anthropogenic emissions of CH_4 and N_2O, respectively). Land use change contributes 17.4% of global GHG emissions. The largest producer is power generation at 25.9% followed by industry with 19.4% (adapted from Solomon *et al.*, 2007).[1]

growth in agricultural emissions during the 20th century, a trend which is predicted to continue during the first part of the current century.[1] Agriculture, like many other sectors of activity, is faced with demands to reduce emissions of GHGs in line with national, regional and global policies and agreements, but faces a dilemma. The global human population is growing rapidly and is expected to reach 9 billion by 2050 and increased demand for food is predicted from both population increase and changing diets.[2] The agricultural sector is therefore faced with a growing demand for food and the consequent need to increase food supply. However, simultaneously, there has been a surge in the production of crops for biofuels that also increases the pressure on land resources. Additionally, over the same period it is anticipated that climate change will increase the need to use deteriorating environments in which to grow crops and produce livestock. The challenge is therefore to increase agricultural production at the same time as decreasing its burden on the environment, which includes decreasing GHG emissions. This chapter outlines the sources of GHG emissions from agriculture, and describes opportunities for emission reductions through implementation of various mitigation measures.

Agriculture plays a complex role in contributing to GHG emissions. Nitrous oxide (N_2O) emissions from fertilisers and manures and methane (CH_4) emissions from livestock (ruminants and manure management) and rice cultivation are the two principal sources of GHG emissions from the agricultural sector. Net emissions of carbon dioxide (CO_2) from agriculture are relatively small, although land use change (for example conversion of forested land to cropland) can make very large contributions to emissions. The land-use sector is, however, unusual in that as well as releasing CO_2, it is able to remove it from the atmosphere. It is estimated that since preindustrial times, around one third of emissions of CO_2 to the atmosphere have originated from land use change.[3] Over this period, soils used in agricultural production have lost organic carbon (OC), a natural constituent, and one which is important in contributing to soil fertility. However, the process of carbon loss is reversible, and measures can be introduced to remove CO_2 from the atmosphere and sequester it in soils, creating benefits both in terms of climate and of agricultural production.[4]

The non-CO_2 GHGs produced by agriculture vary significantly by region. The Asian continent is the most significant source, contributing to 1670 Mt CO_2 equivalents (CO_2eq) in 2005. By contrast, Western Europe and the Organisation for Economic Co-operation and Development (OECD) countries of North America were each responsible for less than 600 Mt CO_2eq in the same year. The Middle East and North Africa have the highest projected growth rates, of 95%, in agricultural GHG emissions between 1990 and 2020. Significant increases are also expected in South East Asia where emissions are already high, but lower increases of 18% are expected in OECD countries and North America.[5]

Emissions of the non-CO_2 GHGs have grown significantly since preindustrial times. N_2O concentrations have risen from preindustrial concentrations of 270 ppbv to around 323 ppbv in 2010,[6] with 70% of current emissions attributable to agriculture. The term Global Warming Potential (GWP) is used as a measure of the net warming effect on our atmosphere of GHG over a fixed period of time relative to that of CO_2, which is assigned a GWP of one. GWPs are expressed per unit mass of a substance. N_2O is a particularly potent GHG with a GWP 296 times greater than that of CO_2, over a 100 year period.[1] The corresponding GWP of CH_4 is 25 times greater than that of CO_2. The accumulation of GHGs in our atmosphere that has occurred as a result of human activity is considered to have caused recent changes to our climate, and threatens to cause further significant climate change. In the longer term (towards the end of this century) this is likely to have negative effects on the sustainability of agriculture at a global scale.[1]

2 Current Agricultural Sources of Nitrous Oxide, Methane and Carbon Dioxide

Since the beginnings of industrialisation and the rapid growth of human populations all three biogenic GHGs (N_2O, CH_4 and CO_2) have shown rapid growth in concentrations in our atmosphere (Figure 2), contributing significantly to the warming or radiative forcing of the planet (contributions from increasing concentrations of N_2O, CH_4 and CO_2 to radiative forcing being approximately 4%, 17% and 61%, respectively).[7] The activities attributable to this rise in concentrations are varied, but agriculture and land use change (also see Chapter 1) have been important in contributing to emissions of each of these gases.

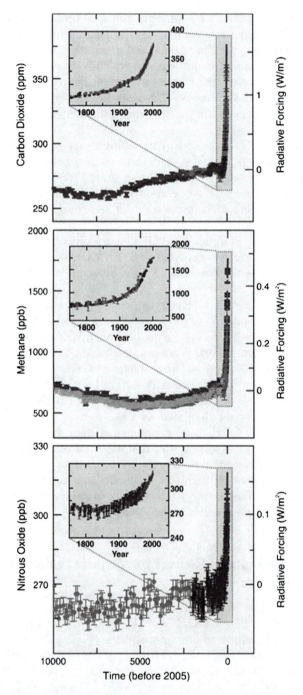

Figure 2   Growth in atmospheric concentrations of the greenhouse gases: carbon
dioxide, methane and nitrous oxide over the past 10 000 years (large panels)
and since 1750 (inset panels). The corresponding radiative forcings relative to
1750 are shown on the right hand axes of the large panels (from IPCC 2007).[7]

2.1 Nitrous Oxide

Nitrous oxide is one of the stable GHGs, with an atmospheric lifetime of approximately 120 years. This has the consequence that emissions from current day activities will have long-term implications even if immediate action is taken to curb emissions. A further environmental concern about N_2O is that it also contributes to the destruction of stratospheric ozone (O_3).[8] N_2O is converted to nitric oxide (NO) by reaction with excited stage oxygen atoms mainly in the stratosphere, and the subsequent reactions of NO create a catalytic cycle of O_3 destruction. The reaction between NO (from N_2O) and O_3 is a natural mechanism contributing to the removal of N_2O from our atmosphere, but increased N_2O concentrations are causing this process to proceed more rapidly, leading to O_3 depletion. The presence of O_3 in the stratosphere is important because it provides protection to living organisms from harmful UV radiation.

Although N_2O is a naturally occurring gas, current atmospheric concentrations of around 323 ppb are 16% above preindustrial levels, with an increase of about 0.26% per year. The main cause of the increase in atmospheric N_2O has been anthropogenic emissions from agriculture (Figure 3).[9] These are principally in the form of reactive nitrogen (Nr) derived from nitrogen fertilisers, livestock manures, urine deposited by grazing livestock, biological nitrogen fixation (BNF), mineralised soil organic matter (SOM) from cultivating old pasture or forest, and biomass burning.

N_2O is generated by the microbial processes of nitrification and denitrification. Nitrification involves the oxidation of ammonium nitrogen to nitrate. Two groups of microorganisms are involved: 'autotrophs', organisms that

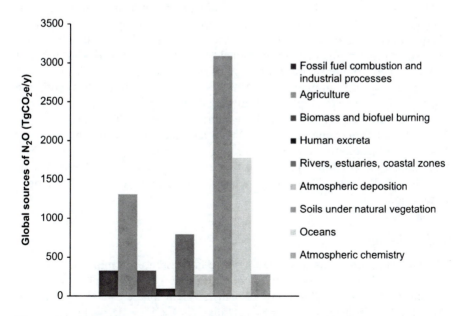

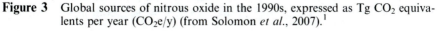

Figure 3 Global sources of nitrous oxide in the 1990s, expressed as Tg CO_2 equivalents per year (CO_2e/y) (from Solomon *et al.*, 2007).[1]

derive energy from sunlight or chemical reactions (in this case the nitrification process), and 'heterotrophs' that derive their energy from the breakdown of organic matter (OM) sources. Autotrophic nitrification involves two stages as shown in Equation (1), in which ammonium is first oxidised *via* the intermediate hydroxylamine to nitrite, and is then further oxidised to nitrate.[10]

$$N_2O$$
$$\nearrow \quad \uparrow$$
$$NH_3 \rightarrow NH_2OH \rightarrow NO_2^- \rightarrow NO_3^- \tag{1}$$

These oxidation processes release a small amount of energy on which the nitrifier organisms are dependent. N_2O can be released as a by-product both of the oxidation of ammonium to nitrite, and subsequent oxidation of nitrite to nitrate. The heterotrophic nitrification process is also able to oxidise ammonium to nitrate, but the process is less well understood and is likely to be more common in fungal communities and in more organic soils. However, it is thought to be a less important process than that of autotrophic nitrification.[10]

Denitrification (whether heterotrophic or autotrophic) is a reductive sequence of reactions converting nitrate to dinitrogen gas *via* intermediates including the gases NO and N_2O, as shown in Equation (2). A wide spectrum of microbial groups is capable of undertaking denitrification.

$$NO_3^- \rightarrow NO_2^- \rightarrow NO \rightarrow N_2O \rightarrow N_2 \tag{2}$$

Denitrification can also occur as a result of chemodenitrification (a non-biological reduction of nitrate). This happens as a consequence of abiotic chemical decomposition and can take place in sterile soils; although it is difficult to distinguish this process in the field from biological processes, it is not thought to contribute significantly to N_2O emissions.[11]

One key aspect of mitigating N_2O emissions from agriculture is quantifying all the various nitrogen inputs so that the additional nitrogen required as fertiliser on crops and grassland can be accurately calculated. This is not easy for BNF and nitrogen from mineralised OM (often called 'Soil Nitrogen Supply' or SNS); quantification is improving for manures, organic materials[12] and crop residues, but is very difficult for atmospheric deposition; fortunately the latter is usually small.[13] Quantifying these inputs is critical to optimising nitrogen applications.

2.2   Methane

Methane concentrations in the atmosphere have risen even more steeply than those of N_2O, from preindustrial concentrations of ~700 ppbv to today's concentrations of ~1800 ppbv (Northern Hemisphere ~1850 ppbv and Southern Hemisphere ~1750 ppbv).[6] Unlike N_2O, CH_4 is derived from a wide variety of sources including agriculture, gas exploration, combustion, mining,

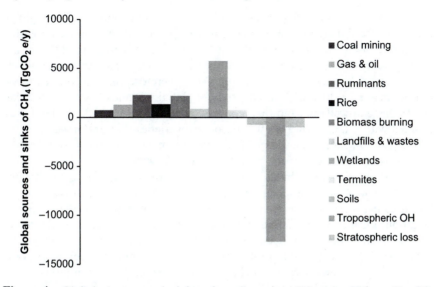

Figure 4 Global sources and sinks of methane in 1999, expressed as Tg CO_2 equivalents per year (CO_2e/y) (from Solomon *et al.*, 2007).[1]

and waste (Figure 4).[1] Agriculture is responsible for approximately 37% of net anthropogenic CH_4 emissions of current global CH_4 emissions.[14] The main sink for atmospheric CH_4 is oxidation by reaction with hydroxyl radicals in the atmosphere. CH_4 has an atmospheric lifetime of approximately 12 years, making it a good target for mitigation, given that the benefits of reduced emissions would be apparent over a relatively short time.

Most CH_4, whether it be from fossil fuel or a more contemporary source, is produced as a result of biologically mediated reduction processes (*e.g.* methanogenesis), in which OM is used by microorganisms as a terminal electron acceptor in the process of respiration. The two best described pathways involve the use of CO_2 and acetic acid as terminal electron acceptors, as shown in Equations (3) and (4).

$$CO_2 + 4H_2 \rightarrow CH_4 + 2H_2O \qquad (3)$$

$$CH_3COOH \rightarrow CH_4 + CO_2 \qquad (4)$$

Methanogenesis occurs in a wide variety of highly reduced environments including natural and agricultural wetlands, the gut of ruminant livestock, manure stores and landfill sites, but the underlying microbiology is the same.

Unlike N_2O, CH_4 originates from a wide range of sources which include both biogenic and industrial ones. Natural and cultivated wetlands are the single most important source, but others include livestock farming, biomass burning, fossil fuels, landfill sites and termites.[15] Approximately 60% of global emissions are derived from human activity; within agriculture, rice cultivation and livestock management are particularly important, contributing 25% of global emissions. Recent studies have indicated that the growth rate in accumulation of atmospheric CH_4 has slowed almost to zero.[16] The reasons for this are not entirely clear. Changing rates of atmospheric consumption are thought unlikely to be responsible and it has been suggested therefore that various emission sources may be reducing. A recent study of global wetlands has suggested that the drainage of wetlands and their conversion to various agricultural land uses, coupled with a move towards the use of rice varieties that are higher yielding and depend on shorter periods of flooding, may have contributed significantly to reductions in the emissions of CH_4 at a global scale.[16]

Agriculture also contributes significantly to global CH_4 emissions through the management of livestock and their manures.[1] At the centre of issues about how livestock-based CH_4 is measured, managed and mitigated, is the process of digestion. Farmed land animals fall broadly into two categories: ruminants or non-ruminants, the latter being principally pigs and poultry. In 2009, there were 1.382 billion cattle, 0.867 billion goats and 1.071 billion sheep worldwide. Other CH_4 producing livestock included 188 million buffalo, 25 million camels, 8 million other camelids and 21 million horses, mules and donkeys.[14] Ruminants degrade fibrous carbohydrate feeds due to the presence of symbiotic microorganisms in the rumen.[17] The main end-products of ruminal fermentation include volatile fatty acids (VFA), CO_2, and hydrogen. VFAs are a source of energy and precursors to synthetic processes that the animal can utilise. As hydrogen is poisonous for the microorganisms that conduct the process, the animal uses its rumen to manage this by generating CH_4 from other methanogenic bacteria. This gas is not usable for the animal so it passes out by periodic eructation, or is absorbed into the bloodstream through the gut wall and expelled continually in breathing through the lungs.

Ruminants typically consume vast amounts of grass or conserved fodder, grazing for 8 to 12 hours per day and ruminating for much of the balance. In addition, they may receive a proportion of their diets through higher energy concentrated food sources, some from primary agricultural grain and protein production from all over the world, but often waste and by-products from other parts of the human food chain. In some systems, all feed can come from these non-grass sources. Feed type, its nutritive value and level of intake, will indirectly determine the amount of CH_4 produced by influencing microorganisms in the gut. Cellulolytic bacteria degrade cellulose and their derivates (or non-soluble carbohydrates). High fibre-content diets, with lower digestibility will produce ratios of VFA with more hydrogen production, and they spend longer within the rumen, reducing feed intake. The higher the fibre content, broadly the higher levels of CH_4 produced. As the maturity stage of the vegetation increases, its fibre content increases and hence CH_4 emissions tend to increase. The very thing that the ruminant is good at, converting poor quality grassland into human food, becomes a major climate change issue for the areas of the world with poor quality forage, namely rangelands and tropical systems where grazed, cut and conserved vegetation for these systems have lower digestibilities than that used in more intensive systems in temperate zones. To further compound these factors, many extensive and small scale producer systems have lower output productivity, such as lower growth rates and lower reproductive rates.[17]

2.3 Carbon Dioxide

Carbon dioxide is released largely from microbial decay or burning of plant litter and SOM.[18] Humans have drastically altered the global carbon cycle, mostly through increased use of fossil fuels and land use change (Figure 5).[19] On a global level, agricultural land accounts for 37% of the Earth's land surface,[20] and because of their large extent and use, agricultural lands have a significant impact on the global carbon cycle. The global soil organic carbon (SOC) inventory is estimated to be 1500 Pg, which is three times the amount of carbon stored in vegetation and twice the amount in the atmosphere.[21] Global estimates of historic soil carbon losses through cultivation and disturbance are in the range of 40–90 Pg carbon, with current rates of carbon loss due to land use change of about 1.6 ± 0.8 Pg C y^{-1}, mainly in the tropics.[22] SOC levels are determined by the balance of net OM inputs (*e.g.* crop residues, organic amendments) and net losses of carbon from the soil through OM decomposition, dissolved OC and loss through erosion (also see Chapter 2). Carbon inputs to the soil are largely determined by the land use and the conversion of native ecosystems to agriculture almost invariably results in a net loss of soil carbon (see Table 1, Section 4.3). The rapid increase in the world's agricultural area over the past 300 years was responsible for large CO_2 emissions in the past

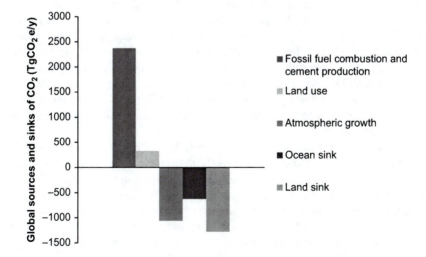

Figure 5 Global sources and sinks of carbon dioxide in 2008, expressed as Tg CO_2 equivalents per year (CO_2e/y). Note that a residual value of ~ 300 TgCO_2/y, which represents the sum of all sources and sinks, is attributed to unaccounted variability in land models and uncertainties in land use change (from Le Quéré *et al.*, 2009).[19]

but because these agricultural soils are now relatively carbon-depleted, they represent a potential CO_2 sink if part of the lost carbon can be regained.[23]

3   International Reporting of Greenhouse Gas Emissions

The procedures that are generally used to report national emissions of GHGs are those described by the Intergovernmental Panel on Climate Change (IPCC).[24] They identify individual activities contributing to sources and sinks of emissions and methods by which these are aggregated and reported. By definition an emission factor (EF) is the rate of emission per unit of activity, output or input.[24] The IPCC inventory methodology is based on a bottom-up aggregation of emissions from different sources: the product of an EF for each source, multiplied by the so-called "activity data". Thus for example direct soil emissions of N_2O are calculated by multiplying the fraction of the applied nitrogen that is emitted as N_2O (EF_1), by the total mass of nitrogen applied. "Default values" for EF_1 are used where there are no measurement data for the country involved that meet the necessary quality criteria. EF_1 was adjusted from 1.25% to 1.00% in 2006, but the former value remains in force for submissions to United Nations Framework Convention on Climate Change (UNFCCC) until 2015.

Different levels or Tiers of precision in the reporting of GHG emissions are recognised by the IPCC. At its most basic level, countries can use internationally agreed EFs that apply across large regions (temperate, tropical *etc.*) of the globe. Thus for example, CH_4 emissions from livestock are calculated by multiplying the number of animals in a particular country by the EF for a particular livestock type and climatic region. Where there is a country-specific value to replace the default value, it may apply to all, or only some, emissions. Thus it is possible to "mix and match", so that, for example, a country might have good enough data for dairy cattle to use its own ("Tier 2") EF, but retain the default for sheep. Where detailed process-based and spatially-referenced modelling approaches are undertaken, it may be possible to report emissions at the most detailed "Tier 3" level. However, this option is not currently in widespread use, due to the model requirements for detailed input data and rigorous validation.

EFs used in Tier 1 calculations are often based on data from temperate and/or humid conditions. However, it is likely that these will need to be revised as emerging evidence from other regions and climate zones, in particular from China, Asia, and South America becomes available. Recent evidence presented at an IPCC experts meeting in 2010 suggested that the default EF_1 for N_2O may be too high for croplands in many world regions. In many areas of the world, the availability of underlying activity data also constrains the ability to report at higher Tiers.[25] The main weakness of the IPCC approach is that, once an EF is established, the emission varies only in response to very crude management change (*e.g.* fertiliser use or livestock numbers), particularly in a Tier 1 calculation. There is a strong case for moving on from here; scientific evidence supports the important role of regional climate (in particular of the gradients from arid to humid and from oceanic to continental), livestock breed characteristics, crop type, *etc*. Statistical and process-based models can help in estimating local EFs and in stratifying agricultural systems by typical relations between driving factors and GHG emissions, but the inclusion of management practices for large-scale estimates remains difficult.

3.1 Indirect Emissions – Is there a Gap between Top-down and Bottom-up Global Budgets?

Using a global top-down approach, taking account of the steady increase in the N_2O concentration in the atmosphere (corresponding to 3.9 Tg N_2O-N yr^{-1}), Crutzen *et al.* (2008)[26] estimated that 3–5 per cent of all new Nr input into terrestrial systems is converted to N_2O. This N_2O EF was based on data compiled by Prather *et al.* (2001)[27] and Galloway *et al.* (2004)[9]. New Nr input was taken to include fertiliser nitrogen produced by the Haber–Bosch process and BNF. Furthermore, it was assumed that emissions from animal manures and crop residues all emanated from these primary inputs. Based on data from the same sources, Crutzen *et al.* (2008)[26] also showed that this EF was similar to that for terrestrial ecosystems in preindustrial times, when the nitrogen inputs were from natural sources: mainly BNF, plus NOx from lightning.

The IPCC bottom-up approach[24] now provides a default EF_1 for direct emissions of 1%. The defaults for nitrogen loss by volatilization (10% of fertiliser nitrogen, 20% of manure nitrogen), combined with the EF_4 (1%) for N_2O from this nitrogen once redeposited, are equivalent to another 0.1–0.2% of the nitrogen applied to the field. Similarly, the default leaching fraction of 30%, combined with the indirect EF_5 (0.75%), adds a further 0.23%. Thus in total, the default fraction of applied nitrogen that is emitted as N_2O adds up to 1.33–1.43%. This is only one-third of the mean EF of 4% obtained *via* the Crutzen *et al.*[26] approach. However, modelling by Del Grosso *et al.* (2006)[28] based on the "bottom-up" IPCC (2006) methodology gave good agreement with the Crutzen *et al.* method at the global scale: 5.8 Tg N_2O-N yr^{-1}, *via* IPCC (2006),[24] compared with 4.2–7.0 Tg N_2O-N yr^{-1} *via* Crutzen *et al.*[26] The agreement was achieved by the inclusion in the bottom-up procedure of N_2O from the recycling of Nr in animal management systems. Thus, there is convergence of two radically different approaches. Del Grosso *et al.* commented: " ....as scale increases, so does the agreement between estimates based on soil surface measurements (bottom-up approach) and estimates derived from changes in atmospheric concentration of N_2O (top-down approach).". Indeed, the convergence " ....increases confidence in emissions estimates because the methods are based on different assumptions....". Davidson (2009)[29] argued that the Crutzen *et al.*[26] EF underestimated atmospheric accumulation of N_2O emissions in the first half of the 20th century – a period when N_2O concentrations were increasing faster than production of new Nr. He suggested that nitrogen "mined" from newly tilled soils had been a major source of new Nr, and modelled the atmospheric N_2O increase on the basis of

EFs for manure production and synthetic fertiliser-nitrogen of 2.0% and 2.5%, respectively. However, recent calculations[30] show that if the Crutzen *et al.*[26] concept of newly fixed Nr is broadened to include NOx deposition and the Nr mined from hitherto virgin land, then the application of a simple 4% EF to these nitrogen inputs gives a close fit to the observed trend in atmospheric concentration. Although the global estimates can be reconciled, there is an implication that regional matching of top-down and bottom-up may not be as close, depending on whether the dominant agricultural system in the region is representative of the global mix or conversely skewed towards dominance by arable or livestock farming.

The downward revision of indirect emissions due to leaching in *the IPCC 2006 Guidelines*[24] was based on the availability of peer-reviewed studies, and it seems that these were located mostly in temperate zones, with small (short) rivers. These environments are not globally representative, and bigger fluxes may be coming from hot countries/big slow-moving rivers/major deltas, in particular tropical rivers with high OC loads, estuaries and anoxic high-nitrogen offshore zones to which the leached nitrogen is finally transported. There is a need for new methods (*e.g.* aircraft/balloon/satellite-based) to target these regions, making measurements at the necessary regional scale, to confirm or refute their importance in the global emissions total.

Significant progress has been made over the past decade in quantifying CH_4 emissions by use of satellite observation. Recent observations by the Sciamachy spectrometer on board the Envisat satellite have provided important information on the distribution of global CH_4 sources, particularly in regions of the world where terrestrial sources are poorly monitored.[31] These studies have provided detailed spatial and temporal data on emissions from wetland rice cultivation in South East Asia and emissions from tropical Africa and South America which help improve our ability to put together a realistic global budget of CH_4 emissions. These observations also help to identify discrepancies in ground-based observations. Recent European satellite observations have suggested that CH_4 emissions across the North and West of Europe may be up to 40% higher than those reported to the UNFCCC, indicating that the EFs used for the construction of such inventories may be too low.[32] However, because of the large uncertainties associated with such estimates it is possible that the two estimates may be compatible.

Carbon accounting methods used to account for GHG mitigation effectiveness of land use change are based on absolute measurements at a point in time; others take into account the time dimension of carbon sequestration and storage.[24] Emissions and removals of CO_2 within the agriculture, forestry and other land use sector are generally estimated on the basis of changes in ecosystem carbon stocks. Changes in carbon stocks may be estimated by direct inventory methods or by process models. For example, changes in SOC due to land management/land use change can be estimated using an approach which uses SOC change on a per area basis under land use and management change derived from long term experiments and scenarios of land use change to estimate total change in SOC for a whole area (see Table 1, Section 4.3).[33,34]

An estimate of the flux of CO_2 to the atmosphere from land use change of 1.6 (0.5 to 2.7) Gt C yr^{-1} was found for the 1990s, based on a combination of techniques, but much uncertainty in the net CO_2 emissions due to land use change remains.[7] One of the highest uncertainties is attributable to the net effect of changing soil erosion through land management because an unknown portion of eroded carbon is stored in buried sediments of wetlands, lakes, river deltas and coastal zones.[35]

4 Future Mitigation Strategies

4.1 Nitrous Oxide

As noted earlier in the chapter, N_2O is mostly produced by the microbially-mediated soil processes of nitrification and denitrification. These are a natural part of the nitrogen cycle and essential for life on Earth. Nitrification is a step in the mineralisation of organic nitrogen to nitrate and a key process in the life cycle of all organisms: life – death – decay and decomposition, providing the basic nutrients for life. N_2O is a by-product of nitrification, see Equation (1), so the aim is to limit N_2O production but not wholly prevent nitrification. Denitrification is the reduction of nitrate through to dinitrogen, N_2O being the penultimate product in the process, see Equation (2). Denitrification is the only process that returns Nr back to dinitrogen and so closes the nitrogen cycle.[9] The increasing amounts of Nr in the environment are caused by the fixation of nitrogen for fertiliser, biomass burning, cultivation and the growth of legumes, which "natural" denitrification cannot compensate for. Again, therefore, we do not want to stop denitrification but to develop management techniques that drive the process through to completion to dinitrogen when and where it occurs.

The process controls of both nitrification and denitrification are the availability of nitrogen and carbon, temperature and moisture and oxygen concentration (aerobicity): nitrification requires an aerobic environment; denitrification an anaerobic/anoxic environment. However, it has been shown that, due to the spatial heterogeneity of soils, both processes can occur simultaneously, even in the same soil aggregate.[36] Recent research has shown that soil physicochemical properties (bulk density, OM, OC, N and C : N ratio) have an overriding influence on the potential denitrification activity.[37] Maintaining high rates of soil nitrogen and OC over a long period support the development of microbial communities that produce more N_2O when conditions are conducive to denitrification, compared to soils containing less nitrogen and OC. There are also structural differences in denitrifier communities in soils with high nitrogen and carbon contents: they possess proportionally fewer copies of the N_2O reductase gene, nosZ, so may be less able to close the nitrogen cycle by reducing N_2O to N_2.[37] This reinforces the need to avoid excessive inputs of nitrogen, whether as fertiliser, manure or atmospheric deposition, and of OC. Nevertheless, this must be balanced with the need for adequate SOM to maintain soil structure and reduce energy use in soil management (tillage), which of themselves reduce GHG emissions.

4.1.1 Optimising Nitrogen Use by Crop Plants.   As Figure 6 shows, losses of nitrogen as N_2O increase exponentially above the optimum for crop growth in absolute terms and when weighted for yield.[38] There is also evidence of a non-linear relationship between nitrogen fertiliser application rate and N_2O emissions from grazed grassland.[39] The most straightforward

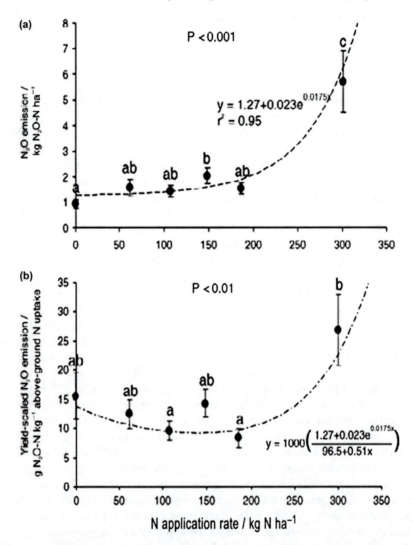

Figure 6 N_2O emissions a) and yield-scaled N_2O emissions b) from arable crops (maize, wheat, potatoes, flooded rice) graphed against N application rates (from van Groenigen *et al.*, 2010).[38]

mitigation option is therefore matching the nitrogen application, whether as fertiliser, manure or BNF, to the crop requirements both in space and time. Many papers and advisory guides have been written to facilitate this.[12,40] Soil heterogeneity and the dependence of SNS and crop growth on the weather make it a complex process. However, recent research[41] has found that even simple methods for predicting SNS and fertiliser requirements, such as the Field Assessment Method in the UK fertiliser Manual[12] are beneficial when the recommendation tables are used correctly. The additional benefit of reducing fertiliser nitrogen applications, through improved nitrogen use efficiency from the range of nitrogen sources (SNS, manures, BNF), should be that less fertiliser nitrogen would need to be manufactured, thus reducing the CO_2eq associated with its production.

4.1.2 Optimising Nitrogen Use by Livestock. The efficiency of nitrogen utilization in ruminants is very variable at between 10 and 40% but generally poor with a mean of *ca.* 25%. Low efficiency means large amounts of nitrogen are excreted in urine and faeces, making efficient recycling of these 'wastes' key to minimising emissions. Efforts to reduce nitrogen losses *via* urine/faeces, and associated N_2O emissions, have focused on understanding the key mechanisms involved in the control of nitrogen metabolism, including the efficiency of nitrogen capture in the rumen and the modification of protein degradation through modifying the feed (physically and chemically).[42,43]

4.1.3 Inhibitors. Although it is not sensible or practicable to try and stop nitrification and denitrification, various inhibitors have been used to delay nitrification with the aim of retaining nitrogen in the form of ammonium or, for urea, of preventing its hydrolysis. This can reduce N_2O losses during nitrification (if the ammonium is taken up by the plant), direct denitrification of nitrate and denitrification of leached nitrate.[43] Recent developments in the use of inhibitors include their incorporation into the fertiliser prill or granule, or the spraying of the inhibitor onto grazed pasture when the stock are moved off, which is proving effective.[43,44] The costs of inhibitors has tended to restrict their use, but recent increases in the price of fertiliser nitrogen makes them an economic option and legally enforced restrictions on nitrogen losses would also increase take-up.

4.1.4 Soil Management and Tillage. The main focus of better soil management in the context of N_2O emissions is better structure to avoid waterlogging and compaction and to increase aeration, thus limiting denitrification.[45] This could, of course, increase nitrification and N_2O emission by that process. However, as the largest emissions of N_2O are usually produced by denitrification, and good soil structure has other benefits (better root and so crop growth, less energy needed in cultivation, better water infiltration and so less erosion), good soil structure is important. Drainage of heavy soils could contribute to this, but it is likely to lead to greater leaching losses and indirect N_2O emissions. Maintaining the correct soil pH is important in agriculture; advice is that it should be *ca.* 6.0–6.2 for grassland and 6.5–7.0 for arable land.[12] However, the effects of soil pH on N_2O emissions are complex: absolute rates of denitrification decrease with pH but the ratio of N_2O to N_2 increases.[46]

Zero- or Minimum-tillage are often advocated for improved soil management, particularly OM content, water holding capacity and structure. Johnson *et al.* (2007)[47] found that Zero- or Minimum-tillage can lead to greater bulk density and increased denitrification and N_2O emissions. Ussiri *et al.* (2009)[48] found that long-term Zero-Tillage reduces N_2O emissions; Oorts *et al.* (2007)[49] found the opposite. Possibly over time, soil structure under Zero- or Minimum-tillage stabilises and earthworms become more active and develop a good, open structure. It would appear that short-term or rotational Zero- or Minimum-tillage, as widely practised in the UK, is likely to increase N_2O emissions; long-term Zero- or Minimum-tillage may be a mitigation option but one that requires conclusive proof.

4.1.5 Land Use Change. The cultivation of organic soils results in a large amount of SOM being mineralised each year, with N_2O being produced during nitrification and the denitrification of surplus nitrate over crop requirements.[50] Land use change in terms of stopping cultivation of these soils is therefore likely to reduce N_2O emissions, but they are some of the most fertile and productive soils in the world. With food security an issue, ceasing production does not appear to be an option (see also Section 4.5). The Royal Society has expressed a need for 'Sustainable Intensification'.[51] Whilst this could increase inputs of nitrogen, the aim is 'More for Less' including better nitrogen use efficiency. As Van Groenigen *et al.* (2011; Figure 6)[38] have shown, this should lead to reduced emissions per unit of product. Given the pressing need to feed >9 billion people by 2050, this is surely an important approach to follow.

4.2 Methane

4.2.1 Methane from Ruminant Livestock. Many studies have considered mitigation options, both within a single farming system, but also at a national/international level. Smith (2007)[5] summarised three main routes to mitigation: through standard animal feed changes, specific additives or treatments, and system efficiencies. Many mitigation strategies are likely to be cumulative, some are competitive, and many need combinations of animal, feed and management to make them practical and effective in practice.

4.2.2 Dietary Opportunities. A range of options is potentially available to manage CH_4 production within the rumen, either controlling methanogenesis *per se* or shifting other nutritional components within the rumen. Vaccination, adding new micro-organisms directly, feeding tannins or other plant secondary compounds are amongst potential opportunities. Increasing the level of the oil within the diet has been demonstrated to reduce CH_4 output, but mainly with high, and arguably impractical, levels of oil inclusion. Ionophores, particularly the monensin supplement Posilac$^©$ (Monsanto Inc.), are widely used in practice in many feedlot based systems in North America.[52] This product is one of a number of potentially mitigating routes banned in the European Union. As described in Section 2.2, improving diet quality whether by improving forage quality or by using higher energy non-forage feeds typically reduces methanogenic output. For example within growing livestock systems, an increase in energy intake through higher energy value per kg of feed will increase growth rate. But to compound this effect, with greater digestibility, feed intake increases. Gross feed energy intake further increases. This in turn leads to greater growth rate (which ultimately becomes weight of output) which is created for marginally lower increases in CH_4 output. The goal of improved production through increased feed quality is a realistic goal for many producers for economic reasons, but is often constrained by the environment and by the genotype of the animals involved. Converting land-based and often self-sufficient systems to systems that 'import' feed to create feedlot type systems is highly debateable. Such systems have shifted pastoral agriculture with high values for landscape and biodiversity to village-based feedlot systems with local environmental challenges and loss of traditional grazing. Other direct techniques to positively modify production, such as improved feed efficiency through use of growth hormones (*e.g.* bovine somatotropins) or through direct growth-stimulating hormones, are widely in use in some parts of the world but are also banned within the European Union.[53]

4.2.3 Avoiding Inefficiencies. Animals with greater efficiency produce more milk and meat per unit of intake, and per unit of CH_4 output. Choosing, and breeding from, animals with greater resource efficiency are already implicit in many breed improvement programmes. There are numerous reports of these programmes improving the net GHG emissions.[54] Animal diseases create direct losses (deaths and failure of the milk or meat product to be marketable) as well as losses in efficiency (decline in productivity and failure to digest and use nutrients efficiently). It is widely recognised that improving health, survival and progress to market will all improve the ratio of product to CH_4 output. Much of animal disease and loss is avoidable or at worst can be reduced by improved practices. There are many treatments and management approaches which are judged impractical or not cost-effective, but they are effective and there is capacity to go beyond current good practice to reduce this inefficiency.

Improvement procedures in herd or flock efficiency are highly relevant. These include reducing replacement rate, increasing longevity, improving reproductive efficiency and in cattle lowering the age that heifers achieve their first calving, as fewer unproductive animals are kept in the herd. Increasing productive animal longevity is seen as a valid means to reduce flock/herd overhead costs. This requires targeting of management practices that remove or reduce the reason that otherwise productive animals are culled *e.g.* for infertility in dairy cows or teeth loss in sheep. These are most likely to come from a combination of both management and targeted breed improvement.[52]

4.2.4 Livestock Reduction or Replacements. Removing the CH_4 producing animal is the ultimate CH_4 mitigation and a number of bodies and authors promote this by calling for a change in land use, which must be driven by either government action or more likely through a change in market demand for meat and dairy products. For example, Zerocarbonbritain[55] proposes a reduction of the current area used for food-based agriculture by 71%, with biomass production and forestry using much of the released land, and with the grazing animal virtually extinct. Such a shift would involve radical changes to diets (to *ca.* 25% of current meat intakes), difficult to imagine in a market-based economy.

4.2.5 Methane from Wetland Rice. Methane emissions from agricultural wetlands occur mostly as a result of emissions from paddy rice production. The most important factor controlling these emissions is the period of flooding, which determines the soil's redox potential and potential CH_4 production. In recent years there has been a move away from permanently flooded production systems, to more productive cultivation systems involving periodic flooding or dryland production. Water management that involves multiple flooding and draining of the soil has been shown to reduce CH_4 emissions by 40%.[52] Crop residue management and fertiliser nitrogen management can also affect CH_4 emissions. A reduction of crop residue addition prior to flooding can reduce emissions as can the use of nitrate based fertilisers. However, both of these management approaches have tradeoffs in relation to net GHG emissions. Where residue incorporation is reduced, the soil will accumulate less carbon in the SOM pool. In circumstances where nitrate fertilisers are applied, there can be significant loss of N_2O, as microorganisms switch from OM to nitrates as the terminal electron acceptor.[52]

4.3 Carbon Dioxide

Soil carbon sequestration is the building of soil carbon sinks, and in agricultural soils this can be achieved through increasing carbon inputs and storing a larger proportion of the carbon from net primary productivity in the longer term soil carbon pools, or by slowing down decomposition.[56] Improving the productivity and sustainability of existing agricultural lands is crucial to help reduce the rate of new land clearing, from which large amounts of CO_2 from biomass and soil are emitted to the atmosphere.[23] For managed agricultural land, best management practices that increase carbon inputs to the soil (*e.g.* improved residue and manure management) or reduce losses of carbon from soil decomposition and erosion (*e.g.* reduced impact tillage, reduced residue removal) help to maintain or increase SOC levels. Also, agro-forestry systems increase both the standing stock of carbon in the aboveground biomass and can also enhance soil carbon sequestration when compared with equivalent areas without trees.[57] It should be remembered that soil carbon sinks resulting from sequestration activities are not permanent and will continue only for as long as appropriate management practices are maintained. If a land management or land use change is reversed, the carbon accumulated will be lost, usually more rapidly than it was accumulated.[58] Efficient carbon sequestration in agricultural soils demands a permanent management change and implementation concepts adjusted to local soil, climate and management features in order to allow selection of areas with high carbon sequestering potential.[33]

Smith (2004)[57] estimated that soil carbon sequestration could meet at most about one third of the current yearly increase in atmospheric CO_2-C, but the duration of the effect would be limited, with significant impacts lasting only 20–50 years. Furthermore, with the population growing and diets changing in developing countries, more land is likely to be required for agriculture.[22] Smith *et al.* (2010)[34] calculated the per-area mitigation potentials for land use change in the UK (Table 1). It can be seen that SOC tends to be lost when converting grasslands or forests to croplands and that SOC tends to increase when restoring grasslands, forests or native vegetation on former croplands. In addition to these land use changes, draining, cultivating or liming highly

Table 1 Mean estimates of per-area mitigation potentials for land use change on mineral and organo-mineral soils expressed in t CO_2-eq ha^{-1} y^{-1} (from Smith *et al.*, 2010).[34]

Land-use change	Mean estimate CO_2 (t CO_2-eq ha^{-1} y^{-1})[a]	Mean estimate for all GHGs (t CO_2-eq ha^{-1} y^{-1})[a,b]
Permanent grass to cropland	−9.50	−9.50
Permanent grass to temporary grass	−7.85	−7.85
Permanent grass to forestry	−2.25	−2.25
Cropland to permanent grass	3.04	5.34
Cropland to temporary grass	0.30	0.53
Cropland to forestry	1.59	1.59
Temporary grass to permanent grass	7.85	7.85
Temporary grass to cropland	−1.54	−1.54
Temporary grass to forestry	1.59	1.59
Forestry to permanent grassland	1.54	1.54
Forestry to cropland	−6.16	−6.16
Forestry to temporary grass	−6.16	−6.16

[a]Positive values represent reduced emissions or enhanced removal; negative values represent increased emissions or suppressed removal.
[b]For mineral soils, there were insufficient data for estimating changes in baseline methane emission/oxidation under land use change, but for cropland to permanent grass and cropland to temporary grass and land use changes, there was sufficient information to estimate changes in baseline nitrous oxide after land use changes.

organic soils can also result in SOC losses but restoration of organic soils to their native condition results in SOC gains.[22]

Similarly, Freibauer et al. (2004)[33] identified the most promising measures for soil carbon sequestration in the agricultural soils of Europe to be: (i) the promotion of organic inputs on arable land instead of grassland, (ii) the introduction of perennials (grasses, trees) on arable set-aside land for conservation or biofuel purposes, (iii) to promote organic farming, (iv) to raise the water table in farmed peatland, and (v) zero tillage or conservation tillage, but some options posed potential risks of increasing N_2O emissions.

4.4 Greenhouse Gas Mitigation Potential: Combined Effects of all Gases

Much research has focused on CO_2-C mitigation and has largely ignored potential effects of land management change on the important agricultural GHGs: CH_4 and N_2O. Often a practice will affect more than one gas, by more than one mechanism, sometimes in opposite ways, so that the net benefit depends on the combined effects of all gases,[20] as illustrated by calculated GHG mitigation potentials in Table 1. Schulze et al. (2009)[59] reviewed recent estimates of European CO_2, CH_4 and N_2O fluxes between 2000 and 2005, using both top-down estimates based on atmospheric observations and bottom-up estimates derived from ground-based measurements. Both methods yield similar fluxes of GHGs, suggesting that CH_4 emissions from livestock and N_2O emissions from arable agriculture are fully compensated for by the CO_2 sink provided by forests and grasslands. Comparison between the carbon and the GHG balance of continental Europe showed that current land management reduces the terrestrial GHG sink, which could otherwise offset non-biological GHG emissions. As a result, the balance for all GHGs across Europe's terrestrial biosphere is near-neutral, despite carbon sequestration in forests and grasslands. Worryingly, the increasing trends towards more intensive agriculture and more intensive forest management (not solely a European issue, as land use changes at a greater scale have already been seen and are likely in the future on other continents) are likely to make Europe's land surface a significant source of GHGs. Overall, findings from this continental-scale GHG balance study indicated that the development of land management policies which aim to reduce GHG emissions should be a priority.

Recent studies on large-scale land use change within the agricultural sector[34] have shown GHG mitigation potentials are an order of magnitude greater than GHG emissions from changes in land management practice (*ca.* 243 Mt CO_2eq per 20 yr or 12 Mt CO_2eq per year on an annual basis). However, this land use strategy is likely to be a limited option for GHG mitigation as land use is largely determined by market factors for agricultural products and the climatic suitability of the land.[34] Therefore, much of the potential for mitigation in the agricultural sector will be within the limits of changing management practices on land that remains in the same agricultural use.[57]

4.5 The Economics of Mitigation

Some of the mitigation options discussed above are available now, but adoption of different measures is likely to be influenced by costs and time of delivery. Cost curve analysis is a useful tool for assessing the economics of practicability and provides a rational basis for ranking mitigation measures on the basis of cost and carbon savings.[60,61] Marginal abatement cost curves are particularly useful because they can help policy makers to set targets for GHG mitigation in agriculture.[61] Cost curve analysis of global GHG mitigation options by McKinsey (2010; Figure 7)[60] indicated that better nutrient and residue management will both result in small reductions in GHG emissions but also save a significant amount of money compared to the business as usual case. Compared to both measures, better rice management has a smaller cost saving per tonne of GHG abated but a larger potential for total abatement.[60]

Better nutrient and residue management means optimising nutrient, especially nitrogen, inputs; the aim is to match crop requirements to nitrogen inputs as closely as possible, thus resulting in less nitrogen at risk of denitrification (and leaching) (see Section 4.1.1).[40] This option would also include the use of nitrification inhibitors. Management practices that result in significant GHG abatements all cost the farmer money: reducing pastureland conversion and better grassland management generally have a large potential for abatement and cost relatively little; restoring organic soils would cost a little more but be an effective abatement, but probably more for CO_2; degraded

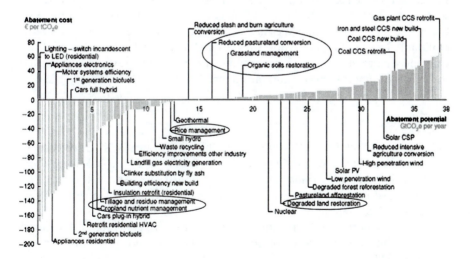

Figure 7 A cost–curve analysis of global GHG mitigation options in CO_2 equivalents (CO_2e). Bars below the zero line on the x axis represent a net cost saving compared to the "business as usual" case, whilst those above this line represent a net cost increase. Agricultural options are marked with ellipses (from McKinsey, 2010).[60]

land restoration would cost a significant amount and gain relatively little. Economically then, the best current mitigation options are better nutrient, tillage and residue management, together with the avoidance of converting pasture into arable land.

5 Conclusions

It is clear that GHG emissions from agriculture represent a major contribution to global GHG budgets. In terms of each component gas CO_2, N_2O and CH_4 and their contribution to agricultural emissions, N_2O and CH_4 are deemed the most important in agriculture because of their greater contributions. Mitigation technologies and practices include reducing emissions, enhancing removals (*e.g.* soil carbon sequestration) and avoiding (or displacing) emissions. Approaches that best reduce emissions depend on local conditions and therefore vary from region to region across the globe. There is scope within the agricultural sector to further reduce GHG emissions and to increase soil carbon sinks but measures to reduce emissions of GHGs have to be considered within a global context and international co-operation is required. Furthermore, as many mitigation practices affect more than one GHG, it is important to assess the impact of all GHGs simultaneously. The development of future mitigation strategies needs to take account of the compelling need to maintain or increase food production as discussed. We described mitigation options that are relevant to individual GHGs, however, it should be emphasised that mitigation should aim to reduce overall GHG emissions, and where a measure results simply in a different partitioning of emissions, it will be less valuable. Furthermore, GHG mitigation measures should not increase emissions of other pollutants (*e.g.* ammonia). Future research should not be centred mainly on CO_2 and studies of land–atmosphere interactions should be extended to include the effects of land use and land management on the GHG balance.

References

1. S. Solomon, D. Qin, M. Manning, Z Chen, M. Marquis, K. B. Averyt, M. Tignor and H. L. Miller, *Climate Change 2007: The Physical Science Basis*, Contribution of Working Group I to the Fourth Assessment Report of the Intergovernmental Panel on Climate Change, Cambridge University Press, Cambridge, 2007.
2. H. C. J. Godfray, J. R. Beddington, I. R. Crute, L. Haddad, D. Lawrence, J. F. Muir, J. Pretty, S. Robinson, S. M. Thomas and C. Toulmin, *Science*, 2010, **327**, 812.
3. J. A. Foley, R. DeFries, G. P. Asner, C. Barford, G. Bonan, S. R. Carpenter, F. S. Chapin, M. T. Coe, G. C. Daily, H. K. Gibbs, J. H. Helkowski, T. Holloway, E. A. Howard, C. J. Kucharik, C. Monfreda, J. A. Patz, I. C. Prentice, N. Ramankutty and P. K. Snyder, *Science*, 2005, **309**, 570.
4. R. Lal, F. Follett, B. A. Stewart and J. M. Kimble, *Soil Sci.*, 2007, **172**, 943.

5. P. Smith, D. Martino, Z. C. Cai, D. Gwary, H. Janzen, P. Kumar, B. McCarl, S. Ogle, F. O'Mara and C. Rice, *Agric. Ecosyst. Environ.*, 2007, **118**, 6.
6. Advanced Global Atmospheric Gases Experiment (AGAGE), 2011; http://agage.eas.gatech.edu/data.htm (accessed 20/12/2011).
7. R. K. Pachauri and A. Reisinger, *Climate Change 2007: Synthesis Report*, Contribution of Working Groups I, II and III to the Fourth Assessment Report of the Intergovernmental Panel on Climate Change, IPCC, Geneva, 2007.
8. A. E. Waibel, T. Peter, K. S. Carslaw, H. Oelhaf, G. Wetzel, P. J. Crutzen, U. Poschl, A. Tsias, E. Reimer and H. Fischer, *Science*, 1999, **283**, 2064.
9. J. N. Galloway, F. J. Dentener, D. G. Capone, E. W. Boyer, R. W. Howarth, S. P. Seitzinger, G. P. Asner, C. C. Cleveland, P. A. Green, E. A. Holland, D. M. Karl, A. F. Michaels, J. H. Porter, A. R. Townsend and C. J. Vorosmarty, *Biogeochemistry*, 2004, **70**, 153–226.
10. N. Wrage, G. L. Velthof, M. L. van Beusichem and O. Oenema, *Soil Biol. Biochem.*, 2001, **33**, 1723–1732.
11. O. van Cleemput and L. Baert, *Plant Soil*, 1984, **76**, 233.
12. Department for Environment, Food and Rural Affairs (DEFRA), *The Fertiliser Manual (Reference Book 209)*, The Stationery Office, London, 2010.
13. *Review of Transboundary Air Pollution (RoTAP)*, DEFRA; http://www.rotap.ceh.ac.uk/documents, 2011 (accessed 20/12/2011).
14. Food and Agriculture Organisation, *The State of Food and Agriculture: Livestock in the Balance*, FAO, Rome, 2009; http://www.fao.org/docrep/012/i0680e/i0680e00.htm (accessed 20/12/2011).
15. European Commission Joint Research Centre (JRC) and The Netherlands Environmental Assessment Agency (PBL), Emission Database for Global Atmospheric Research (EDGAR), The Netherlands, 2011; http://edgar.jrc.ec.europa.eu/index.php (accessed 20 December 2011).
16. M. Heimann, *Nature*, 2011, **476**, 157–158.
17. S. Tamminga, A. Bannink, J. Dijkstra and R. Zom, *Feeding Strategies to Reduce Methane Loss in Cattle*, Report. 34, Animal Science Group, Wageningen, 2007.
18. H. H. Janzen, *Agric. Ecosyst. Environ*, 2004, **104**, 399.
19. C. Le Quéré, M. R. Raupach, J. G. Canadell, G. Marland, L. Bopp, P. Ciais, T. J. Conway, S. C. Doney, R. A. Feely, P. Foster, P. Friedlingstein, K. Gurney, R. A. Houghton, J. I. House, C. Huntingford, P. E. Levy, M. R. Lomas, J. Majkut, N. Metzl, J. P. Ometto, G. P. Peters, C. Prentice, J. T. Randerson, S. W. Running, J. L. Sarmiento, U. Schuster, S. Sitch, T. Takahashi, N. Viovy, G. R. van der Werf and F. I. Woodward, *Nature Geos.*, 2009, **2**, 1.
20. P. Smith, D. Martino, Z. Cai, D. Gwary, H. H. Janzen, P. Kumar, B. McCarl, S. Ogle, F. O'Mara, C. Rice, R. J. Scholes, O. Sirotenko, M. Howden, T. McAllister, G. Pan, V. Romanenkov, U. Schneider,

S. Towprayoon, M. Wattenbach and J. U. Smith, *Trans. R. Soc.*, 2008, **363**, 789.

21. Intergovernmental Panel on Climate Change, Special Report on Land Use, Land Use Change, and Forestry, Cambridge University Press, Cambridge, 2000.
22. P. Smith, *Nutr. Cycling Agroecosyst.*, 2008, **81**, 169.
23. K. Paustian, O. Andrén, H. H. Janzen, R. Lal, P. Smith, G. Tian, H. Tiessen, M. Van Noordwijk and P. L. Woomer, *Soil Use Manage.*, 1997, **13**, 230.
24. *Intergovernmental Panel on Climate Change Guidelines for National Greenhouse Gas Inventories*, Prepared by the National Greenhouse Gas Inventories Programme, IPCC, Japan, 2006.
25. E. Lokupitiya and K. Paustian, *J. Environ. Qual.*, 2006, **35**, 1413.
26. P. J. Crutzen, A. R. Mosier, K. A. Smith and W. Winiwarter, *Atmos. Chem. Phys.*, 2008, **8**, 389.
27. M. Prather, D. Ehhalt, F. Dentener, R. Derwent, E. Dlugokencky, E. Holland, I. Isaksen, J. Katima, V. Kirchhoff, P. Matson, P Midgley, M. Wang, T. Berntsen, I. Bey, G. Brasseur, L. Buja, W. J. Collins, J. Daniel, W. B. DeMore, N. Derek, R. Dickerson, D. Etheridge, J. Feichter, P. Fraser, R. Friedl, J. Fuglestvedt, M. Gauss, L. Grenfell, A. Grübler, N. Harris, D. Hauglustaine, L. Horowitz, C. Jackman, D. Jacob, L. Jaeglé, A. Jain, M. Kanakidou, S. Karlsdottir, M. Ko, M. Kurylo, M. Lawrence, J. A. Logan, M. Manning, D. Mauzerall, J. McConnell, L. Mickley, S. Montzka, J. F. Müller, J. Olivier, K. Pickering, G. Pitari, G. J. Roelofs, H. Rogers, B. Rognerud, S. Smith, S. Solomon, J. Staehelin, P. Steele, D. Stevenson, J. Sundet, A. Thompson, M. van Weele, R. von Kuhlmann, Y. Wang, D. Weisenstein, T. Wigley, O. Wild, D. Wuebbles and R. Yantosca, in *Atmospheric Chemistry and Greenhouse Gases in Climate Change 2001: The Scientific Basis*, ed. J. T. Houghton, Y. Ding, D. J. Griggs, M. Noguer, P. J. van der Linden, X. Dai, K. Maskell and C. A. Johnson, Cambridge University Press, Cambridge, 2001.
28. S. Del Grosso, W. Parton, A. Mosier, M. Walsh, D. Ojima and P. Thornton, *J. Environ. Qual.*, 2006, **35**, 1451.
29. E. A. Davidson, *Nature Geosci.*, 2009, **2**, 659.
30. K. A. Smith, A. R. Mosier, P. J. Crutzen and W. Winiwarter, *Philos. Trans. R. Soc. London, Ser. B.*, in press.
31. P. Bergamaschi, C. Frankenberg, J. F. Meirink, M. Krol, M. G. Villani, S. Houweling, F. Dentener, E. J. Dlugokencky, J. B. Miller, L. V. Gatti, A. Engel and I. Levin, *J. Geophys. Res. Atmos.*, 2009, 114.
32. P. Bergamaschi, M. Krol, J. Meirink, F. Dentener, A. Segers, J. van Aardenne, S. Monni, A. Vermeulen, M. Schmidt, M. Ramonet, C. Yver, F. Meinhardt, E. Nisbet, R. Fisher, S. O'Doherty. and E. Dlugokencky, *J. Geophys. Res. Atmos.*, 2010, 115.
33. A. Freibauer, M. Rounsevell, P. Smith and A. Verhagen, *Geoderma*, 2004, **122**, 1.

34. P. Smith, A. Bhogal, P. Edgington, H. Black, A. Lilly, D. Barraclough, F. Worrall, J. Hillier and G. Merrington, *Soil Use Manage.*, 2010, **26**, 381.
35. P. Smith, K. W. T. Goulding, K. A. Smith, D. S. Powlson, J. U. Smith, P. D. Falloon and K. Coleman, *Nutr. Cycling Agroecosyst.*, 2001, **60**, 237.
36. K. Khalil, B. Mary and P. Renault, *Soil Biol. Biochem.*, 2004, **36**, 687–699.
37. I. M. Clark, N. Buchkina, D. Jhurreea, K. W. T. Goulding and P. R. Hirsch, *Philos. Trans. R. Soc. London, Ser. B.*, in press.
38. J. van Groenigen, G. Velthof, O. Oenema, K. Van Groenigen and C. van Kessel, *Eur. J. Soil Sci.*, 2010, **61**, 903.
39. L. M. Cardenas, R. Thorman, N. Ashlee, M. Butler, D. Chadwick, B. Chambers, S. Cuttle, N. Donovan, H. Kingston, S. Lane and D. Scholefield, *Agric. Ecosyst. Environ.*, 2010, **136**, 218.
40. K. W. T. Goulding, S. Jarvis and A. Whitmore, *Philos. Trans. R. Soc. London, Ser. B.*, 2008, **363**, 667.
41. D. Kindred, S. Knight and P. Berry, *Establishing Best Practice for Estimation of Soil N Supply*, Final report of Project 3425, HGCA, Stoneleigh, 2011.
42. S. Calsamiglia, A. Ferret, C. Reynolds, N. Kristensen and A. van Vuuren, *Animal*, 2010, **4**, 1184.
43. J. Luo, C. de Klein, S. Ledgard and S. Saggar, *Agric. Ecosyst. Environ.*, 2010, **136**, 282.
44. S. Chien, L. I. Prochnow and H. Cantarella, *Adv. Agron.*, 2009, **102**, 267.
45. I. M. Young and K. Ritz, *Soil Tillage Res.*, 2000, **53**, 201.
46. M. Heinen, *Geoderma*, 2006, **133**, 444.
47. J. M. F. Johnson, A. J. Franzluebbers, S. L. Weyers and D. C. Reicosky, *Environ. Pollut.*, 2007, **150**, 107.
48. D. A. Ussiri, R. Lal and M. K. Jarecki, *Soil Tillage Res.*, 2009, **104**, 247.
49. K. Oorts, R. Merckx, E. Grehan, J. Labreuche and B. Nicolardot, *Soil Tillage Res.*, 2007, **95**, 133.
50. C. L. van Beek, M. Pleijter, C. M. J. Jacobs, G. L. Velthof, J. W. van Groenigen and P. J. Kuikman, *Nutr. Cycling Agroecosyst.*, 2010, **86**, 331.
51. Royal Society, *Reaping the Benefits: Science and the Sustainable Intensification of Global Agriculture*, The Royal Society, London, 2009.
52. P. Smith, D. Martino, Z. Cai, D. Gwary, H. Janzen, P. Kumar, B. McCarl, S. Ogle, F. O'Mara, C. Rice, B. Scholes and O. Sirotenko, Agriculture, in *Climate Change 2007: Mitigation of Climate Change, Contribution of Working Group III to the Fourth Assessment Report of the Intergovernmental Panel on Climate Change*, ed. B. Metz, O. R. Davidson, P. R. Bosch, R. Dave and L. A. Meyer, Cambridge University Press, Cambridge, UK and New York, NY, USA, 2007.
53. F. P. O'Mara, *Animal Feed Sci. Technol.*, 2011, **166–167**, 7.
54. G. C. Waghorn and R. S. Hegarty, *Animal Feed Sci. Technol.*, 2011, **166–167**, 291.
55. Zero Carbon Britain 2030, Centre for Alternative Technology; 2010: http://www.zerocarbonbritain.com/ (accessed 20/12/2011).
56. P. Smith, *Soil Use Manage.*, 2004, **20**, 212.

57. N. Fitton, C. P. Ejerenwa, A. Bhogal, P. Edgington, H. Black, A. Lilly, D. Barraclough, F. Worrall, J. Hillier and P. Smith, *Soil Use Manage.*, 2011, **27**, 491.
58. P. Smith, D. S. Powlson and M. J. Glendining, 1996. Establishing a European soil organic matter network (SOMNET), in *Evaluation of Soil Organic Matter Models using Existing, Long-term Datasets*, NATO ASI Series, ed. I, D. S. Powlson, P. Smith and J. U. Smith, Springer-Verlag, Berlin, 1996, **38**, 81.
59. E. D. Schulze, S. Luyssaert, P. Ciais, A. Freibauer and I. A. Janssens, *Nature Geosci.*, 2009, **2**, 842.
60. McKinsey, *Impact of the Financial Crisis on Carbon Economics*, Version 2.1 of the Global Greenhouse Gas Abatement Cost Curve, McKinsey & Company, 2010.
61. D. Moran, M. MacLeod, E. Wall, V. Eory, A. McVittie, A. Barnes, R. Rees, C. F. E. Topp and A. Moxey, *J. Agric. Econ.*, 2011, **62**, 93.

Impacts of Agriculture on Water-borne Pathogens

DAVID KAY,* JOHN CROWTHER, CHERYL DAVIES,
TONY EDWARDS, LORNA FEWTRELL, CAROL FRANCIS,
CHRIS KAY, ADRIAN MCDONALD, CARL STAPLETON,
JOHN WATKINS AND MARK WYER

ABSTRACT

Microbial indicators of water quality are used to quantify the risk derived from faecally contaminated surface and drinking waters. The historical focus in this area has centred on human-derived sewage contamination of bathing, shellfish and drinking waters. However, emerging catchment-scale water legislation in North America and Europe, in particular, is driving a more holistic approach in which quantification of microbial pollution from all sources is undertaken, to inform and prioritise appropriate remedial action designed to ensure health risk is minimised. This involves integrated management of agricultural livestock-derived pollution alongside sewage effluents to ensure compliance of impacted sites with appropriate regulatory standards. The evidence-base for the design of best management practices by farmers which will remove and/or attenuate microbial flux from catchment systems is very limited when compared to the chemical parameters associated with ecological impairments, such as phosphorus and nitrogen. However, early empirical investigations do suggest the potential to realise very significant water quality benefits from simple interventions, such as stock exclusion fencing of stream banks and well-designed constructed wetland systems. Further process-based investigation of these areas is underway and this research effort is becoming imperative as emerging experience of catchment-scale legislation strongly suggests the importance of microbial pollution as the

*Corresponding author

Issues in Environmental Science and Technology, 34
Environmental Impacts of Modern Agriculture
Edited by R.E. Hester and R.M. Harrison
© The Royal Society of Chemistry 2012
Published by the Royal Society of Chemistry, www.rsc.org

principal reason for non-compliance with water quality standards in
North America.

1 Introduction

This chapter describes the principal sources and flow paths of microbial pollution from farming activities to water courses and the available management
options for attenuation of this pollutant flux. The evaluation is set into a policy
framework to contextualise the relevance of catchment microbial dynamics in
agricultural systems.

There are four principal sources of microbial pathogen loadings to rivers and
coastal waters, namely:

1. Human sewage disposal systems which discharge *via* pipes known as
   'point-source' discharges. These can be further split into:
   a. 'Continuous discharges' of sewage effluent following various levels of
      treatment from simple screening of solids (*e.g.* rags and plastics) to full
      biological treatment and disinfection. The receiving waters can include
      rivers, lakes or coastal waters through outfall pipes; and
   b. 'Intermittent discharges' which occur when the sewerage system is
      receiving urban surface water drainage after rainfall. This can
      increases the flow beyond the capacity of the sewer, thus, producing
      short-term discharges of untreated, but diluted, raw sewage to
      rivers, lakes and coastal waters. This is normally discharged
      through:
      i. 'Combined sewer overflows' in the sewer line itself,
      ii. 'Storm tank overflows', generally sited just upstream of waste
          water treatment works, and
      iii. 'Pumping station overflows', which operate when a pumping
           station on the sewer line is overwhelmed because of rainfall-
           supplemented flow volume. These overflows may also operate
           during dry weather conditions in response to emergency conditions
           such as pumping station failure or sewer blockages.
2. Livestock-derived microbial fluxes deriving from a range of catchment
   sources commonly termed 'diffuse-source pollution'. The pattern of this
   loading is site specific but includes:
   a. Voiding of faeces to land (and, in some, cases to water) by livestock at;
      i. In-field feeding and drinking points;
      ii. Gates between fields used for daily stock movements, notably
          for dairy farming operations where stock may be moved
          between milking facilities and fields, particularly in the summer
          period;
      iii. River crossing points which can be in the form of simple fords or
           bridges; and
      iv. Stream bank drinking points, where faeces may be voided
          directly into the watercourse or onto land which has been highly

trampled and 'puddled/poached' by livestock, creating a mobile and available faecal store.

b. Faeces directly voided onto farm yards and other hardstanding areas which stock use during farming activities such as milking. Some of this loading will commonly be scraped and delivered to a dedicated manure or slurry store (see 2c below) but residues remaining after normal yard cleaning will generally be washed from the impervious farm hardstanding area, often entering small drainage ditches and watercourses, particularly where the farm hardstandings have good hydrological connectivity to the catchment drainage network.

c. Livestock manure and slurry, which accumulates from housed livestock, mainly during the winter period in temperate climates, and is commonly stored in slurry tanks or lagoons. This is spread onto fields as a fertilizer and can be timed to maximise feed crop growth. However, sometimes the imperative to spread slurry is driven by the operational requirement to increase available storage capacity after periods of high rainfall, which can encourage inappropriate spreading at high risk periods. Where slurry is applied to fields with good hydrological connectivity to adjacent streams, the microbial loading to the stream environment can be very high, with associated problems of high biochemical oxygen demand from organic-rich material input and high nutrient loadings.

3. Wildlife populations in catchment systems are another potential source of microbial pathogen loading. Generally, microbial loadings from wildlife in intensively farmed catchments are low compared to the livestock and/or human population inputs. However, this area has not received intensive research attention[1] and some surprising findings have been reported, suggesting that roof runoff, commonly perceived to be a clean and sustainable resource, can be highly contaminated with faecal indicator organisms (FIOs), such as intestinal enterococci from avian wildlife sources.[2]

4. Urban diffuse microbial pollution presents a further source. This comprises street drainage and associated drain flow, generally separate from the foul sewerage system, which may be contaminated with faecal matter deposited onto urban roads, pavements and roofs. Canine and feline pets certainly contribute to this loading, as do avian and rodent urban wildlife populations. This is a difficult loading to quantify because many urban areas may also have a number of inappropriate and 'informal' connections of foul sewage from individual properties into the non-foul surface water drainage systems. Links between the foul sewer and surface water drainage networks may also result from ageing and poorly maintained infrastructure. Thus, quantification of the true quality of urban diffuse microbial fluxes is problematical because identification of sites which can be guaranteed free of any such misconnections in the urban upstream infrastructure is extremely difficult.

2 Policy Developments

Until very recently, in Europe at least, there has been no control on microbial loading from sewerage and/or agricultural-related discharges. Discharges of treated sewage effluents have been traditionally 'consented' in the developed nations, based largely on their biochemical oxygen demand and suspended sediment concentrations with additional control of parameters important for fish life such as ammoniacal nitrogen. Taken together, this suite of effluent quality measurements has traditionally been termed 'the sanitary parameters' by the engineering and operational communities designing and managing sewerage networks and waste water treatment works when, in reality, they have no public health (*i.e.* 'sanitary') significance whatsoever. Likewise, regulation of farming activities has generally been directed to maintain the ecological health of streams and surface waters and has tended to focus on biochemical oxygen demand and nutrient loadings from farm-related sources.

This policy environment is, however, changing rapidly in many countries where catchment-focused water management legislation is being implemented. In the United States of America, the *Clean Water Act* (USCWA) represents one of the first such frameworks and it presents a useful lesson charting the priorities emerging after approximately 15 years of implementation.[3] Under the USCWA, where water quality is defined as *impaired*, *i.e.* fails to reach target levels, a 'total maximum daily load' (TMDL) assessment is undertaken to underpin rectification of the impairment. Figure 1 shows the number of identified USCWA impairments from 1996 to 2011.

Some 74 912 water quality 'impairments' were reported between 1st January 1996 and 16th October 2011. Over the same period some 49 064 'total maximum daily loads' were approved by the United States Environmental Protection Agency (USEPA) to address identified impairments. The top five reasons for water quality impairment, subsequently leading to an agreed 'total maximum daily load' investigation, account together for 77.6% of all 'total maximum

Causes of Impairment for 303(d) Listed Waters
Description of this table

NOTE: Click on a cause of impairment (e.g. pathogens) to see the specific state-reported causes that are grouped to make up this category. Click on the "Number of Causes of Impairment Reported" to see a list of waters with that cause of impairment.

Cause of Impairment Group Name	Number of Causes of Impairment Reported
Pathogens	10,981
Metals (other than Mercury)	7,614
Organic Enrichment/Oxygen Depletion	7,074
Nutrients	6,829
Polychlorinated Biphenyls (PCBs)	6,385
Sediment	6,156
Mercury	4,979
pH/Acidity/Caustic Conditions	3,111
Cause Unknown - Impaired Biota	3,589
Turbidity	3,131
Temperature	3,045
Salinity/Total Dissolved Solids/Chlorides/Sulfates	1,899
Pesticides	1,878

Figure 1   United States Clean Water Act identified water quality impairment between 1996 and 2011; http://iaspub.epa.gov/waters10/attains_nation_cy.control?p_report_type=T (accessed 16th October, 2011).

daily loads' to date. In rank order, using the USEPA categories, these are: microbial pollution by faecal indicator organisms (FIOs), incorrectly termed 'pathogens', (10 281 'total maximum daily loads); heavy metal pollution (8168 'total maximum daily loads'); mercury (6992 'total maximum daily loads'); nutrients (6203 'total maximum daily loads'); and sediments and siltation (4403 'total maximum daily loads').[3,4]

It is interesting to note that microbial water quality 'impairments' were the most important reasons for US 'total maximum daily load' studies suggesting a higher US prominence for this area than, for example, nutrients, pesticides and biochemical oxygen demand.

The European Union Water Framework Directive (EUWFD)[5] presents a strikingly parallel set of legislative aims and catchment-scale approaches to those outlined in the USCWA. This Directive is due for implementation in 2015 and European Union (EU) member countries are currently preparing for mandatory compliance. It is the most significant piece of EU water legislation to date and it is interesting to note that it is being applied in a very similar cultural and economic environment to the USCWA, which pre-dates the EUWFD by approximately 20 years. In the EUWFD 'non-compliance' with the EUWFD and/or its daughter Directive standards is equivalent to the USCWA 'impaired water' status and, for such sites, the EUWFD (Article 11) sets out the requirement for a 'programme of measures' involving integrated point and diffuse-source control to ensure compliance in an agreed timescale (*i.e.* equivalent to the USCWA 'total maximum daily load' approach). Thus, the dominance of microbial water quality issues identified through implementation of the USCWA as seen in Figure 1 has significance for the European policy community responsible for implementation of the EUWFD. It is therefore surprising and of concern that, to date, there has been very little policy attention directed to this area when compared to the considerable effort devoted to compliance with nutrient and sediment parameters. This balance has been driven by the policy community focus on the requirement to achieve good ecological status in EU surface waters (see EUWFD, Annex V).[5] This is understandable given the emphasis in the EUWFD Directive text on this new criterion. However, it ignores the requirement, also clearly stated in the EUWFD, that the mechanisms set out in the EUWFD should be used to ensure compliance of 'protected areas' (identified in EUWFD, Annex IV) with water quality standards set out in daughter Directives. The most relevant 'protected areas' (*i.e.* for which microbial pollution can cause proven health risks and/or non-compliance with daughter Directive standards) are those used for bathing and/or harvesting of shellfish, which include filter feeders known to concentrate microbial pathogens in the edible shellfish flesh.[6–8]

The science evidence-base of catchment microbial dynamics, which is essential to inform catchment-scale policies as required by the EUWFD, lags well behind that available for the nutrient and sediment parameters.[9] The extent to which the USCWA experience could be used to inform implementation of the EUWFD was examined by Kay *et al.*[10] who reviewed the operation of microbial 'total maximum daily loads' in California. This review concluded that the US regulatory experience of catchment microbial dynamics, through 'total maximum daily load' assessments, had not, to date, produced operationally useful empirical science or modelling approaches which could be applied in the EU. In effect, many US authorities were defining FIO concentration limits for discharges to streams and coastal waters (in 'total maximum daily load' terminology the 'concentration-based pollutant allocations') which were simply set at the environmental 'receiving water' concentrations required for recreational and shellfish harvesting waters (*i.e.* a geometric mean faecal coliform (FC) concentration in agricultural and surface drainage discharges to tributary streams of <200 cfu 100 mL^{-1} and a 90th percentile for FC in direct discharge to the coastal water of <43 cfu 100 mL^{-1}). Waste water treatment works and boats were required to achieve a FC median value of zero 100 mL^{-1}. The examples chosen did not address the spatial and temporal characteristics of the input fluxes, which are more relevant than 'concentration'. Nor was the feasibility of achieving these criteria addressed. Additionally, sampling programmes were recommended which could not capture data on the hydrological events, which studies world-wide have suggested account for $>90\%$ of the catchment-derived faecal indicator flux from diffuse-source pollution.[11,12]

An additional international policy driver for catchment microbial dynamics research emerged following the publication of the WHO guidelines for drinking water quality.[13,14] These advocated a new regulatory paradigm for drinking water protection based on catchment-scale 'water safety planning'. This mirrors the Hazard Assessment Critical Control Point (HACCP) approaches commonly applied in food processing industries, and more recently applied to bathing beach management and shellfish protection through 'sanitary profiling' of coastal bathing and shellfish harvesting waters.[6,7,15–17]

This development has generated the requirement for water undertakers to assess microbiological risks in water supply gathering grounds. Much of the early research and scientific information in this area has been driven by Australian water undertakers and research scientists and can be traced back to perceived risks from chlorine-resistant protozoan parasites, principally *Cryptosporidium* spp., which have led to studies of both faecal indicator organisms and pathogens in Australian water supply gathering grounds.[18–27] and UK upland supplies.[28] Water undertakers in the EU are now applying similar approaches which imply maintenance of source water quality through the protection of gathering grounds. This is, in effect, a re-invention of the 'multiple barrier' approach to public health protection which had been effectively dismantled, with the introduction of multiple use catchment policies in the developed economies and the consequential heavy reliance on treatment plant and supply system integrity to maintain water quality, over 60 years ago.[29]

3   Microbial Dynamics

3.1   *Pathogens, Indicators and Health Risk*

Microbial standards at regulated water use sites used for bathing, shellfish harvesting or drinking water abstraction are based on faecal indicator bacteria.

These are generally harmless commensal species which are excreted in large numbers and are used to indicate a linkage between faecal contamination and the resource use site. They do not directly indicate pathogen presence at the site. The main reason for this is that pathogens will be present when they are being excreted by the contributing population (both animal and human). Thus, pathogen concentrations in environmental waters tend to vary depending on, often seasonal, patterns of infection and/or illness in the contributing population. Furthermore, there are many hundreds of potential pathogens of interest in any environmental water and absence of one species would certainly not prove the absence of risk from another. For these reasons, the use of microbial indicator species has proven more operationally useful in limiting risk to the potentially impacted population. In effect, the presence of FIOs strongly suggests the existence of faecal contamination which might, at some point, be contributed to by a pathogen carrier. This provides the trigger for management action to limit the contamination, hopefully, before a water user comes into contact with pathogenic species. In addition, the faecal indicator organisms all have simple and quantitative analytical methods which minimise the risk to the analytical laboratory staff associated with culture-based tools for pathogen quantification. It is also important to note that many of the key viral pathogens, such as norovirus, which causes self-limiting gastroenteritis (often termed 'winter vomiting syndrome'), are not 'culturable' and the emerging molecular methods available do not produce the precise enumeration essential for regulatory standards design.[30]

The established faecal indicator organism regulatory parameters used for drinking, bathing and shellfish waters are the coliforms (principally *Escherichia coli*) and the enterococci (principally intestinal enterococci). These are commensal organisms which are a normal part of the mammalian gut flora. Both livestock and human faeces contain very large numbers of these organisms in the 10^6 to 10^8 g^{-1} range (wet weight). There is also considerable variability both between and within species (Table 1) in the daily flux of faecal indicator organisms, with ruminant faeces exhibiting higher concentrations per unit weight than human faeces.

However, using the standard culture methods for enumeration of these FIOs (which employ either growth on culture media after filtration through a membrane to define colony forming units (cfu) or growth in separate culture media after inoculation of defined volumes to define a 'most probable number' (mpn)), the original animal or human source of the organisms, in a resource use site such as a bathing water, is totally indistinguishable. Furthermore, many of the pathogens which might cause infection in the user population can derive from both human and animal faeces. These zoonotic pathogens include the protozoan parasites, *Cryptosporidium* spp. and *Giardia* spp., as well as bacterial pathogens including *E. coli* O157, *Campylobacter* spp. and *Salmonella* spp. The viral pathogens tend to be more species specific. It should be noted, however, that severe and life threatening symptoms can be associated with the zoonotic pathogens, for example *E. coli* O157 is the aetiological agent causing haemolytic ureamic syndrome leading to acute renal failure.

Table 1 Reported *Escherichia coli* burdens associated with key livestock types.

Details	EC burden (cfu animal^{-1} day^{-1})	Source
CATTLE		
Dairy cows	8.99×10^8	(Chambers *et al.*, 2001), (White *et al.*, 2001) and (Muirhead *et al.*, 2005)[a]
Beef cattle	2.54×10^9	Chambers *et al.* (2001) and (Weaver *et al.*, 2005)[a]
Cattle (unclass)	5.40×10^9	(Jones and White, 1984)
	5.00×10^9	Anthony and Morrow (2009)
Median	$\mathbf{3.77 \times 10^9}$	
Calves	2.10×10^{10}	Jones and White (1984)
	3.00×10^{10}	Anthony and Morrow (2009)
Median	$\mathbf{2.55 \times 10^{10}}$	
SHEEP		
Sheep	1.80×10^{10}	Jones and White (1984)
	7.74×10^8	Chambers *et al.* (2001) and Weaver *et al.* (2004)[a]
	2.00×10^9	Anthony and Morrow (2009)
Median	$\mathbf{2.00 \times 10^9}$	
Lambs	$\times 10^{10}$	Chambers *et al.* (2001) and (Vinten *et al.*, 2004)
	7.80×10^9	Anthony and Morrow (2009)
Median	$\mathbf{1.01 \times 10^{10}}$	
OTHER LIVESTOCK		
Pigs	8.90×10^9	Jones and White (1984)
Poultry	2.40×10^8	Jones and White (1984)

[a]Reported in Oliver *et al.* (2009).

Whilst not directly predictive of any specific pathogen's presence, large-scale epidemiological studies have established mathematical relationships between faecal indicator organism presence and illness in water resource users which have formed the evidence-base for standards design. For example, the recent WHO (2003), and resultant EU (Anon, 2006), intestinal enterococci standards for marine bathing waters are based on epidemiological studies using healthy adult volunteers and a randomised controlled trial (RCT) protocol at UK bathing waters.[31–33] This facilitates direct calculation of the health risk associated with a specific water quality standard using a simple probabilistic model of exposure derived from the probability density function of the faecal indicator organism distribution in the bathing waters combined with the calculated dose–response relationship as outlined in Figure 2. This has been published for bathing waters by the WHO in *Guidelines for Safe Recreational Water Environments,* Volume 1, Chapter 4 (Table 2).[6]

Thus, the regulatory pressure on both agricultural and urban microbial discharges is founded on the requirement for faecal indicator organism control

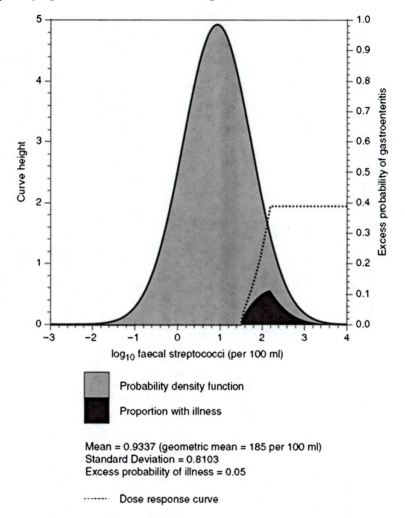

Mean = 0.9337 (geometric mean = 185 per 100 ml)
Standard Deviation = 0.8103
Excess probability of illness = 0.05

········· Dose response curve

Figure 2 The combination of a dose–response curve and probability density function to define the probability of gastrointestinal illness caused by bathing water exposure.
(Source Kay *et al.*, 2004).

leading to numerical compliance with faecal indicator organism standards at water use sites (or 'protected areas' in EUWFD terminology). This compliance requirement is, at least in part, predicated on an evidence-base which defines health risk to water users in terms of faecal indicator organisms, not pathogen, exposures. This evidence-base derives from epidemiological studies, where possible, but often supplemented by quantitative microbial risk assessment (QMRA) which is an established approach to risk management assuming probabilistic distributions of pathogen ingestion, host susceptibility and, sometimes, community transmission of secondary infections.[34–37]

Table 2 Guideline values for microbial quality of recreational waters. (Source: WHO, 2003 and Kay *et al.*, 2004).

95th percentile value of enterococci per 100 mL (rounded values)	Basis of derivation	Estimated risk per exposure
≤40	This value is below the NOAEL in most epidemiological studies.	**<1% GI illness risk** **<0.3% AFRI risk** This relates to an excess illness of less than 1 incidence in every 100 exposures. The AFRI burden would be negligible.
41–200	The 200/100 mL value is above the threshold of illness transmission reported in most epidemiological studies that have attempted to define a NOAEL or LOAEL for GI illness and AFRI.	**1–5% GI illness risk** **>1.9% AFRI illness risk** The upper 95th percentile value of 200 relates to an average probability of one case of gastroenteritis in 20 exposures. The AFRI illness rate at this water quality would be 19 per 1000 exposures, or approximately 1 in 50 exposures.
201–500	This level represents a substantial elevation in the probability of all adverse health outcomes for which dose–response data are available.	**5–10% GI illness risk** **1.9–3.9% AFRI illness risk** This range of 95th percentiles represents a probability of 1 in 10 to 1 in 20 of gastroenteritis for a single exposure. Exposures in this category also suggest a risk of AFRI in the range of 19–39 per 1000 exposures, or a range of approximately 1 in 50 to 1 in 25 exposures.
>500	Above this level, there may be a significant risk of high levels of minor illness transmission.	**>10% GI illness risk** **>3.9% AFRI illness rate** There is a greater than 10% chance of illness per single exposure. The AFRI illness rate at the 95th percentile point of 500 enterococci per 100 mL would be 39 per 1000 exposures, or approximately 1 in 25 exposures.

Notes:

1. Abbreviations used: AFRI = acute febrile respiratory illness; GI = gastrointestinal illness; LOAEL = lowest-observed-adverse-effect level; NOAEL = no-observed-adverse-effect level.

2. The 'exposure' in the key studies (Kay *et al.*, 1994 and Fleisher *et al.*, 1996) was a minimum of 10 min bathing involving three immersions. It is envisaged that this is equivalent to many immersion activities of similar duration, but it may underestimate risk for longer periods of water contact or for activities involving higher risks of water ingestion (see also note 7).

3. The 'estimated risk' refers to the excess risk of illness (relative to a group of non-bathers) among a group of bathers who have been exposed to faecally contaminated recreational water under conditions similar to those in the key studies.

4. The functional form used in the dose–response curve assumes no excess illness outside the range of the data (*i.e.*, at concentrations above 158 intestinal enterococci per 100 mL). Thus, the estimates of illness rate reported above are likely to be underestimates of the actual disease incidence attributable to recreational-water exposure.

5. This table would produce protection of 'healthy adult bathers' exposed to marine waters in temperate north European waters.

6. It does not relate to children, the elderly or immuno-compromised, who would have lower immunity and might require a greater degree of protection. There are no available data with which to quantify this, and no correction factors are therefore applied.

7. Epidemiological data on fresh waters or exposures other than bathing (*e.g.*, high-exposure activities such as surfing, dinghy boat sailing or slalom water canoeing) are currently inadequate to present a parallel analysis for defined reference risks. Thus, a single microbiological value is proposed, *at this time*, for all recreational uses of water, because insufficient evidence exists at present to do otherwise. However, it is recommended that the severity and frequency of exposure encountered by special interest groups (such as bodysurfers, board riders, windsurfers, sub-aqua divers, canoeists and dinghy sailors) be taken into account.

8. Where disinfection is used to reduce the density of indicator bacteria in effluents and discharges, the presumed relationship between intestinal enterococci (as indicators of faecal contamination) and pathogen presence may be altered. This alteration is, at present, poorly understood. In water receiving such effluents and discharges, intestinal enterococci counts may not provide an accurate estimate of the risk of suffering from mild gastrointestinal symptoms or AFRI.

9. Risk attributable to exposure to recreational water is calculated after the method given by Wyer *et al.* (1999), in which a $\log_{10}$ standard deviation of 0.8103 was assumed. If the true standard deviation for a beach were less than 0.8103, then reliance on intestinal enterococci would tend to overestimate the health risk for people exposed above the threshold level, and *vice versa*.

10. Note that the values presented in this table do not take account of health outcomes other than gastroenteritis and AFRI. Where other outcomes are of public health concern, then the risks should be assessed and appropriate action taken.

11. Guideline values should be applied to water used recreationally and at the times of recreational use. This implies care in the design of monitoring programmes to ensure that representative samples are obtained. It also implies that data from periods of high risk may be ignored if effective measures were in place to discourage recreational exposure.

3.2   Catchment Microbial Flux

Quantification of the flux of faecal indicator organisms, and associated pathogens, from a range of catchment sources to impaired resource use sites is an essential first step in the design of evidence-based 'programmes of measures' (EUWFD) or TMDLs (USCWA). In the absence of this information, it would be impossible to assess the relative efficacy and resultant cost effectiveness of alternative remediation strategies, such as microbial attenuation of diffuse-sources driven by agricultural practices, compared with point-source controls, such as sewage effluent disinfection. This type of information is also essential for 'objective' sanitary profiling and/or water safety planning.[6,38]

Two basic foundations of catchment-scale quantitative microbial source apportionment are:

  i. knowledge of the microbial characteristics of sewage effluents; and
  ii. information on microbial fluxes from different land use types covering both agricultural and urban systems.

There is very little peer-reviewed data in this area world-wide when compared to the nutrient parameters of phosphorus and nitrogen. Some data from UK investigations have been published describing faecal indicator organism concentrations in raw sewage and treated effluents,[39] and on the catchment export coefficients from different land use types.[40] The sewage faecal indicator organism concentrations are presented in Table 3 and the export coefficients in Table 4.

Combining the information in Tables 3 and 4 with geographical information system-based land use data, sewage treatment type and population served has been used to provide 'quantitative microbial source apportionment' as part of microbial sanitary profiling for both UK recreational and shellfish harvesting waters.[41–43] Figure 3 shows the predicted riverine FC concentrations produced by this type of 'quantitative microbial source apportionment' modelling in UK catchments draining to the coast in England and Wales, and Figure 4 shows the predicted percentage of the FC flux derived from agriculture.

It is important to note, however, that this modelling approach is dependent on the availability of:

  i. Empirical data on geometric mean faecal indicator organism concentrations under base- and high-flow conditions, river discharge and land use for a range of representative catchments;
  ii. Population served and treatment types for each waste water treatment works in the catchment; and
  iii. The volume and timing of intermittent discharges from combined sewage overflows, pumping station overflows and storm tank overflows in the catchment.

Item iii is highly uncertain in most developed nations where intermittent flows are rarely monitored and modelled data may prove unreliable unless

Table 3 Summary of faecal indicator organism concentrations (cfu 100 mL^{-1}) for different sewage treatment levels and individual types of sewage-related effluents under different flow conditions: geometric means (GMs), 95% confidence intervals (CIs);[a] and results of t-tests comparing base- and high-flow GMs for each group and type;[b] and (in footnote) results of t-tests comparing GMs for the two untreated discharge types and the two tertiary-treated effluent types. (Source Kay et al., 2008b).

Indicator organism / Treatment levels and specific types	Base flow conditions:				High flow conditions:			
	n^c	Geometric mean	Lower 95% CI	Upper 95% CI	n^c	Geometric mean	Lower 95% CI	Upper 95% CI
TOTAL COLIFORMS								
Untreated	253	3.9×10^7 *(+)	3.2×10^7	4.6×10^7	279	8.2×10^6 *(−)	7.0×10^6	9.6×10^6
Crude sewage discharges[d]	253	3.9×10^7 *(+)	3.2×10^7	4.6×10^7	79	1.2×10^7 *(−)	8.2×10^6	1.6×10^7
Storm sewage overflows[d]					200	7.2×10^6	5.9×10^6	8.4×10^6
Primary	130	3.0×10^7 *(+)	2.3×10^7	3.9×10^7	14	1.2×10^7 *(−)	4.0×10^6	3.7×10^7
Primary settled sewage	61	3.8×10^7	3.0×10^7	4.7×10^7	8	2.2×10^7	—	—
Stored settled sewage	26	2.4×10^7	1.2×10^7	5.1×10^7	1	1.1×10^6	—	—
Settled septic tank	43	2.5×10^7	1.3×10^7	4.2×10^7	5	7.5×10^6	—	—
Secondary	853	1.1×10^6	9.5×10^5	1.2×10^6	183	1.3×10^6	1.0×10^6	1.7×10^6
Trickling filter	472	1.4×10^6	1.2×10^6	1.7×10^6	76	1.4×10^6	1.0×10^6	1.9×10^6
Activated sludge	256	7.8×10^5 *(−)	6.2×10^5	1.0×10^6	92	1.4×10^6 *(+)	8.6×10^5	2.1×10^6
Oxidation ditch	35	8.1×10^5	4.6×10^5	1.4×10^6	5	3.1×10^6	—	—
Trickling/sand filter	10	6.4×10^5	2.8×10^5	1.4×10^6	8	2.7×10^5	—	—
Rotating biological contactor	80	6.8×10^5	4.6×10^5	1.0×10^6	2	4.0×10^6	—	—
Tertiary	182	5.5×10^3	3.4×10^3	9.0×10^3	8	3.8×10^3	—	—
Reedbed/grass plot[e]	73	3.7×10^4	1.5×10^4	8.1×10^4	2	2.3×10^4	—	—
Ultraviolet disinfection[e]	109	1.5×10^3	9.9×10^2	2.6×10^3	6	2.1×10^3	—	—
FAECAL COLIFORMS								
Untreated	252	1.7×10^7 *(+)	1.4×10^7	2.0×10^7	282	2.8×10^6 *(−)	2.3×10^6	3.2×10^6
Crude sewage discharges[d]	252	1.7×10^7 *(+)	1.4×10^7	2.0×10^7	79	3.5×10^6 *(−)	2.6×10^6	4.7×10^6
Storm sewage overflows[d]					203	2.5×10^6	2.0×10^6	2.9×10^6

Table 3 Continued.

Indicator organism / Treatment levels and specific types	Base flow conditions:				High flow conditions:			
	n^c	Geometric mean	Lower 95% CI	Upper 95% CI	n^c	Geometric mean	Lower 95% CI	Upper 95% CI
Primary	**127**	$\mathbf{1.0\times10^7}$ *(+)	$\mathbf{8.4\times10^6}$	$\mathbf{1.3\times10^7}$	**14**	$\mathbf{4.6\times10^6}$ *(−)	$\mathbf{2.1\times10^6}$	$\mathbf{1.0\times10^7}$
Primary settled sewage	60	1.8×10^7	1.4×10^7	2.1×10^7	8	5.7×10^6	–	–
Stored settled sewage	25	5.6×10^6	3.2×10^6	9.7×10^6	1	8.0×10^5	–	–
Settled septic tank	42	7.2×10^6	4.4×10^6	1.1×10^7	5	4.8×10^6	–	–
Secondary	**864**	$\mathbf{3.3\times10^5}$ *(−)	$\mathbf{2.9\times10^5}$	$\mathbf{3.7\times10^5}$	**184**	$\mathbf{5.0\times10^5}$ *(+)	$\mathbf{3.7\times10^5}$	$\mathbf{6.8\times10^5}$
Trickling filter	477	4.3×10^5	3.6×10^5	5.0×10^5	76	5.5×10^5	3.8×10^5	8.0×10^5
Activated sludge	261	2.8×10^5 *(−)	2.2×10^5	3.5×10^5	93	5.1×10^5 *(+)	3.1×10^5	8.5×10^5
Oxidation ditch	35	2.0×10^5	1.1×10^5	3.7×10^5	5	5.6×10^5	–	–
Trickling/sand filter	11	2.1×10^5	9.0×10^4	6.0×10^5	8	1.3×10^5	–	–
Rotating biological contactor	80	1.6×10^5	1.1×10^5	2.3×10^5	2	6.7×10^5	–	–
Tertiary	**179**	$\mathbf{1.3\times10^3}$	$\mathbf{7.5\times10^2}$	$\mathbf{2.2\times10^3}$	**8**	$\mathbf{9.1\times10^2}$	–	–
Reedbed/grass plot^e	71	1.3×10^4	5.4×10^3	3.4×10^4	2	1.5×10^4	–	–
Ultraviolet disinfection^e	108	2.8×10^2	1.7×10^2	4.4×10^2	6	3.6×10^2	–	–
ENTEROCOCCI								
Untreated	**254**	$\mathbf{1.9\times10^6}$ *(+)	$\mathbf{1.6\times10^6}$	$\mathbf{2.3\times10^6}$	**280**	$\mathbf{4.9\times10^5}$ *(−)	$\mathbf{4.2\times10^5}$	$\mathbf{5.6\times10^5}$
Crude sewage discharges^d	254	1.9×10^6 *(+)	1.6×10^6	2.3×10^6	79	8.9×10^5 *(−)	6.7×10^5	1.2×10^6
Storm sewage overflows^d					201	3.8×10^5	3.2×10^5	4.5×10^5

	n	GM	GM	GM	n	GM	GM	GM
Primary	**128**	**1.3×10^6**	**1.1×10^6**	**1.7×10^6**	**14**	**9.8×10^5**	**4.4×10^5**	**2.2×10^6**
Primary settled sewage	61	2.4×10^6	2.1×10^6	2.7×10^6	8	1.9×10^6	—	—
Stored settled sewage	26	6.2×10^5	3.2×10^5	1.1×10^6	1	2.9×10^5	—	—
Settled septic tank	41	9.3×10^5	5.3×10^5	1.6×10^6	5	4.3×10^5	—	—
Secondary	**871**	**2.8×10^4 *(−)**	**2.5×10^4**	**3.2×10^4**	**182**	**4.7×10^4 *(+)**	**3.6×10^4**	**6.1×10^4**
Trickling filter	483	4.1×10^4	3.5×10^4	4.7×10^4	76	5.7×10^4	4.2×10^4	8.3×10^4
Activated sludge	262	2.1×10^4 *(−)	1.8×10^4	2.7×10^4	91	4.1×10^4 *(+)	2.7×10^4	6.0×10^4
Oxidation ditch	35	2.0×10^4	1.0×10^4	4.0×10^4	5	1.2×10^5	—	—
Trickling/sand filter	11	2.1×10^4	5.3×10^4	5.3×10^4	8	1.1×10^4	—	—
Rotating biological contactor	80	9.6×10^3	6.7×10^3	1.4×10^4	2	3.7×10^5	—	—
Tertiary	**177**	**3.0×10^2**	**1.8×10^2**	**5.0×10^2**	**8**	**2.1×10^2**	—	—
Reedbed/grass plot[e]	73	1.9×10^3	7.1×10^2	4.3×10^3	2	2.3×10^3	—	—
Ultraviolet disinfection[e]	104	8.3×10^1	4.6×10^1	1.1×10^2	6	9.7×10^1	—	—

[a] CIs only reported where $n \geq 10$.

[b] t-tests comparing low- and high-flow GM concentrations only undertaken where $n \geq 10$ for both sets of samples; only statistically significant ($p < 0.05$) differences between base- and high-flow GM concentrations are reported: indicated by *, with the higher GM being identified as *(+) and the lower value by *(−).

[c] n indicates number of valid enumerations, which in some cases may be less than the actual number of samples.

[d] t-tests comparing the GM concentrations between the two untreated discharge types show high-flow GM concentrations to be significantly higher in crude sewage discharges than storm sewage overflows for TC ($p < 0.05$) and EN ($p < 0.001$).

[e] t-tests comparing the GM concentrations between the two tertiary-treatment effluent types show GM TC, FC and EN concentrations to be significantly higher ($p < 0.001$) in reedbed/grass plot effluents than effluents from UV disinfection for base-flow conditions (there are too few high-flow samples for these tertiary effluents for meaningful comparisons to be made for high-flow GM concentrations).

Table 4 Summary of geometric mean faecal indicator organism (FIO) export coefficients ($\log_{10}$ cfu km^{-2} hr^{-1}) under base- and high-flow conditions at 205 UK sub-catchment sampling points and for various subsets, and results of paired, 1-tailed t-tests to establish whether there are significant elevations at high flow compared with base flow. (Source: Kay et al., 2008b).

FIO		Base flow:			High flow:		
Sub-catchment land use	n	Geometric mean	Lower 95% CI	Upper 95% CI	Geometric mean[a]	Lower 95% CI	Upper 95% CI
TOTAL COLIFORMS							
All subcatchments	205	1.8×10^9	1.4×10^9	2.4×10^9	9.5×10^{10}**	7.2×10^{10}	1.2×10^{11}
Degree of urbanisation[b]							
Urban	20	8.5×10^9	3.3×10^9	2.2×10^{10}	4.1×10^{11}**	1.6×10^{11}	1.1×10^{12}
Semi-urban	60	4.2×10^9	2.6×10^9	6.7×10^9	1.5×10^{11}**	8.3×10^{10}	2.7×10^{11}
Rural	125	9.3×10^8	6.9×10^8	1.3×10^9	6.1×10^{10}**	4.6×10^{10}	8.0×10^{10}
Rural subcatchments with different dominant land uses							
≥75% Improved pasture	15	2.9×10^9	1.4×10^9	6.0×10^9	2.8×10^{11}**	1.6×10^{11}	4.9×10^{11}
≥75% Rough grazing	13	7.1×10^8	3.5×10^8	1.4×10^9	5.3×10^{10}**	2.6×10^{10}	1.1×10^{11}
≥75% Woodland	6	3.1×10^8	5.7×10^7	1.6×10^9	1.4×10^{10}**	6.0×10^9	3.4×10^{10}
FAECAL COLIFORMS							
All subcatchments	205	5.5×10^8	4.1×10^8	7.2×10^8	3.6×10^{10}**	2.7×10^{10}	4.8×10^{10}

Degree of urbanisation[b]						
Urban	20	2.8×10^9	1.1×10^9	7.2×10^9	1.3×10^{11}**	4.8×10^{10}	3.6×10^{11}
Semi-urban	60	1.2×10^9	7.4×10^8	1.9×10^9	4.6×10^{10}**	2.5×10^{10}	8.6×10^{10}
Rural (<2.5% built-up land)	125	2.9×10^8	2.1×10^8	4.0×10^8	2.6×10^{10}**	1.9×10^{10}	3.5×10^{10}
Rural subcatchments with different dominant land uses						
≥75% Improved pasture	15	8.3×10^8	4.3×10^8	1.6×10^9	1.2×10^{11}**	6.5×10^{10}	2.2×10^{11}
≥75% Rough grazing	13	2.5×10^8	1.1×10^8	5.7×10^8	2.5×10^{10}**	1.1×10^{10}	5.5×10^{10}
≥75% Woodland	6	2.0×10^7	4.7×10^6	8.2×10^7	3.3×10^9**	1.3×10^9	8.8×10^9
ENTEROCOCCI						
All subcatchments	205	8.3×10^7	6.6×10^7	1.1×10^8	7.1×10^9**	5.5×10^9	9.3×10^9
Degree of urbanisation[b]						
Urban	20	4.0×10^8	2.1×10^8	7.6×10^8	2.7×10^{10}**	1.1×10^{10}	6.2×10^{10}
Semi-urban	60	1.5×10^8	9.8×10^7	2.2×10^8	1.1×10^{10}**	6.1×10^9	1.9×10^{10}
Rural (<2.5% built-up land)	125	4.9×10^7	3.7×10^7	6.5×10^7	4.7×10^9**	3.5×10^9	6.3×10^9
Rural subcatchments with different dominant land uses						
≥75% Improved pasture	15	9.6×10^7	5.2×10^7	1.8×10^8	2.2×10^{10}**	1.3×10^{10}	3.8×10^{10}
≥75% Rough grazing	13	3.3×10^7	1.2×10^7	9.0×10^7	3.6×10^9***	1.3×10^9	9.7×10^9
≥75% Woodland	6	8.5×10^6	3.8×10^6	1.9×10^7	3.8×10^8***	1.3×10^8	1.1×10^9

[a]Significant elevations in export coefficients at high flow are indicated: ** $p<0.001$.
[b]Degree of urbanisation, categorised according to percentage built-up land: 'Urban' (≥10.0%), 'Semi-urban' (2.5–9.9%) and 'Rural' (<2.5%).

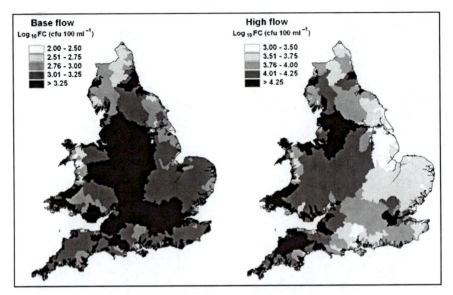

Figure 3 Predicted base- and high-flow GM faecal coliform concentrations (cfu 100 mL^{-1}) in rivers draining to the coast in England and Wales, standardised in base-flow model for a total runoff during the summer bathing season equivalent to 1 mm d^{-1} rainfall, and in high-flow model for high-flow runoff during the summer bathing season equivalent to 1 mm d^{-1} rainfall. (Source: Kay *et al.*, 2010a).

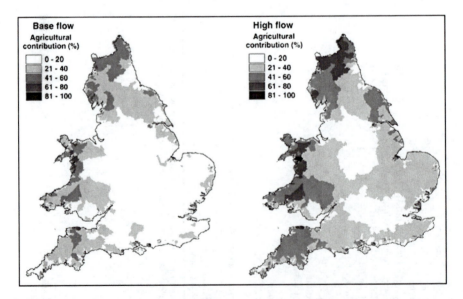

Figure 4 Predicted percentages of GM faecal coliform concentrations in rivers draining to the coast in England and Wales derived from agricultural sources, compared with sewerage-related 'urban' sources under base- and high-flow conditions during the summer bathing season. (Source: Kay *et al.*, 2010a).

validated by long-term empirical measurement. The quantification of inter-mittent fluxes at the catchment scale is therefore a priority area for future investment if sanitary profiling of catchments is to provide a credible evidence-base for the policy community wishing to determine the balance between agricultural and urban microbial flux to protected areas which is essential in the prioritisation of remedial measures in both areas.

3.3 Flux Attenuation and Mitigation of Resource Impacts

The WHO expert group, designing the *2003 Guidelines for Safe Recreational Water Environments*, specifically considered the health risks associated with livestock- and human-derived microbial pollution. The risks attributable to agricultural sources were accepted as an area worthy of further research attention, which has been addressed by a recent WHO-led publication in this area.[44]

The assumption, implicit in the 2003 WHO[6] recreational water guidelines, is that livestock-derived microbial pollution is unlikely to contain high levels of human viral pathogens. Thus, livestock pollution may not generate the levels of minor self-limiting viral gastroenteritis which is the health outcome used in quantifying the risk levels defined in Table 2D.[45] However, the likely presence of zoonotic pathogens would still present a risk which cannot be quantified using the available epidemiological evidence. It was appreciated that this dif-fuse-source risk could not be reduced by simple technical interventions, such as disinfection, as might be the case for a point-source sewage discharge. It was agreed that the nature of this risk was highly episodic and driven by rainfall events which provide the transport energy to move faecally-derived microbial pathogens out of the catchment system to the protected area. This episodic flush process has been observed in empirical investigations world-wide[18,46,47] and it offers an alternative approach to risk mitigation in the case of livestock-derived pathogens which is set out in the WHO guidelines.[6,17]

The WHO approach (Chapter 4)[6] requires:

  i. A sanitary profile to provide a qualitative assessment of whether any episodic pollution loading affecting the resource use area is derived from human or livestock faecal sources;
 ii. That if the pollution is not associated with human sewage, then the managing authorities investigate the possibility of a real-time predictive modelling approach at the exposure location based on previous rainfall or river flow at adjacent real-time monitoring locations;[48–51]
iii. That if the model produces reliable prediction of non-compliance and the managing authority can produce timely warning to the public of adverse water quality (*i.e.* as electronic signs, internet communications and/or hard copy notice boards), then this presents the most effective and practical means of limiting public health risk from episodic agricultural pollution; and
 iv. That given such a predictive management system is in place, any water quality samples collected during periods of an advisory warning to the

public should not be used in calculation of numerical compliance over the full bathing season or bathing seasons (*i.e.* in the case of the WHO Guidelines, this is a 95 percentile value based on 100 samples taken over a 5-year period); *i.e.* such results are, in effect, 'discounted' from the numerical compliance assessment.

The WHO prediction and discounting approach[6] was largely adopted by the European Commission in drafting the EU Bathing Water Directive.[5] There are, however, three key differences. First, there is no EU requirement to demonstrate, through the sanitary survey, that episodic deterioration in water quality is driven only by livestock derived pollution rather than human intermittent sources, such as adjacent combined sewer overflows. Second, the EU Directive allows only 15% of samples to be discounted in any compliance assessment period of 4 years, whereas WHO suggest no such limitation. Third, the EU 4-year period of sampling for compliance assessment implies approximately 16–80 regulatory samples (*i.e.* given the range of sampling options permitted in the EU Directive 2006,[52] rather than the 100 recommended by WHO as the basis for the percentile compliance assessment, the implication being low precision on the percentile calculation for some EU sites with low annual sampling frequencies.

Notwithstanding these differences, the discounting provision in the EU 2006 Directive[52] still offers significant compliance benefits for UK bathing waters which could remain approximately 'compliance-neutral' despite tighter health-based numerical microbial standards in 2015, if discounting is fully implemented. Conversely, if discounting is not implemented, the UK will see significant reductions in bathing water compliance, particularly with the target 'Excellent' standards.[52]

This approach to environmental regulation, of real-time prediction and management action to effect real-time public health protection, is radical and has lessons for many sectors if it can be operationalized as practical management systems. This has been achieved in Scotland by the Scottish Environment Protection Agency (SEPA), the environmental regulator, at 22 demonstration sites (Figure 5) as outlined by Stidson.[49] Further UK work by regulators and water undertakers is now underway to provide demonstration systems in Wales (Swansea Bay) and Yorkshire.

It would be imprudent, however, to assume that episodic agricultural pollution can simply be discounted to effect a 'compliant' bathing water just because the public is warned of consistently poor water quality. The WHO advisory groups considered this scenario and were content that the provision of information to underpin 'informed choice' by the public would result in remedial action if a bathing water was consistently 'unavailable' to the public. It is, therefore, still imperative to seek sustainable attenuation of livestock-derived microbial fluxes at the farm- and field-scale to prevent avoidable impairment of any resource use site. Considerable attention is now being given to the design and evaluation of simple field-scale mitigation measures or 'best management practices' (often termed BMPs) for eliminating and attenuating

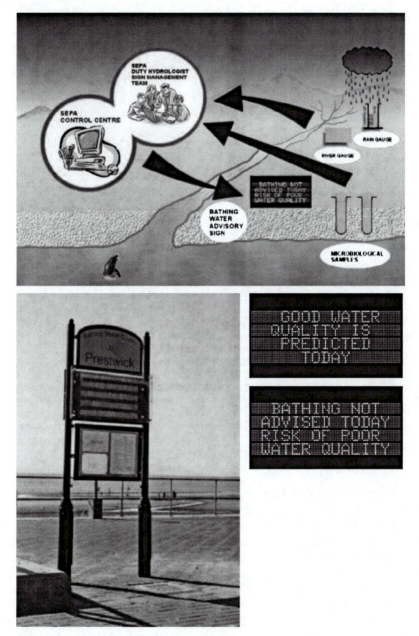

Figure 5 The SEPA bathing water quality prediction system.
(Source: SEPA).

microbial fluxes from agricultural livestock systems. The world-wide literature was recently reviewed for the UK government to inform research and policy in this area.[53] Figure 6 plots the resultant published microbial attenuation data (generally faecal coliform or *E. coli*) produced by the most common 'best

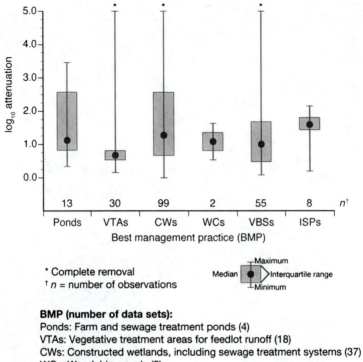

Figure 6 Summary FIO attenuation data for the six BMPs for which datasets have been compiled.
(Source: CREH, 2010).

management practices'. Unfortunately, the effects of farm- and field-scale interventions are often strongly affected by site-specific factors, and are therefore difficult to accommodate in the catchment-scale export coefficient models used for quantitative sanitary profiling or quantitative microbial source apportionment as outlined above.

What is striking about Figure 6 is the extreme range in reported attenuation from almost zero through to 5.0 $\log_{10}$ reduction, which represents almost complete elimination. The principal interventions available are riparian buffer strips, generated easily by stream bank fencing which is set back from the stream edge, and farm ponds/constructed wetlands. The current state of knowledge in these areas has little 'process-based' empirical information available; rather, the studies conducted, to date, provide simple 'black-box' comparative data (*i.e.* input and output) generating the large ranges in Figure 6. Some of this extreme range is explicable by the different monitoring conditions used in different studies. For example, pond systems can produce high attenuation during dry weather conditions, when the retention time is maximised, and outflow from farm hardstandings might be minimal, but have very little attenuation effect following rainfall when the retention time is reduced and high turbidity inputs limit the bactericidal effects of solar irradiance in the pond water column.[40] In the studies used to produce Figure 6, both dry and wet weather periods are monitored and wetlands of various sizes are reported. It is, however, still difficult to identify a clear empirical basis for key wetland design parameters such as: the retention time of complex vegetated wetland systems; and the likely microbial attenuation rates driven by complex processes including solar irradiance, sedimentation and protozoan predation operating in the shallow vegetated environment under variations in flow and seasonality. Simplistic engineering assumptions of a predictable 'reaction-tank' are simply not applicable, especially during the key rainfall-driven periods of high-flow conditions when the attenuation of the wetland is most needed.[2,43] Parallel problems can be identified in the process understanding of riparian buffer strips generated by stream bank fencing. Here, stock exclusion from stream banks has been proven to be effective in reducing high flow microbial flux[54] but, again, very few process data exist to underpin policy-directed design and sizing of the vegetated area and to justify further interventions, beyond stock exclusion from stream banks, to the farming community.

4 Conclusions

Livestock farming can produce significant fluxes of faecal indicator organisms and associated microbial pathogens. Control and management of the former is crucial to mandatory compliance of downstream resource use sites with regulations designed to limit health risks from bathing and shellfish harvesting sites in particular. The regulatory challenge is first to quantify the relative contributions from both urban and agricultural sources of microbial pollution and then to design appropriate and cost-effective remediation strategies to bring non-compliant areas within defined microbial criteria. This objective is, however, challenging for the following reasons:

i. The microbial flux from agricultural systems is highly episodic and is most likely to impair resource use for relatively short periods during and after rainfall events;

ii. Available mitigation measures are at their least effective during such episodic rainfall events which maximise connectivity between catchment microbial 'stores' and the stream network;

iii. Detailed understanding of the catchment dynamics of microbial transport and attenuation, during such events, is very limited;

iv. Emerging regulatory policy accommodates the episodic nature of microbial concentrations in catchment and near-shore systems through the implementation of real-time prediction and discounting, as recommended by WHO (2003) and the EU (2006),[52] thereby offering large potential compliance and public health benefits, but does not replace good farming practice designed to attenuate microbial pollution loadings at source; and

v. Research is needed in this area to parallel the historical attention devoted to other parameters such as the nutrients, the aim being to provide the empirical evidence-base needed to model and predict the effects of field-scale interventions, such as stream bank fencing and constructed wetlands, on high-flow FIO fluxes at the catchment scale, thereby underpinning policy on the design of, and support for, 'best management practice' interventions designed to minimise agricultural loadings and impairment/non-compliance threats to downstream resource use sites.

There is considerable evidence world-wide of increasing attention to microbial catchment loadings and many of these issues are being addressed with the realistic expectation of medium term (5–10 years) delivery of these policy objectives.

References

1. A. T. McDonald, P. J. Chapman and K. Fukasawa, The microbial status of waters in a protected wilderness area, *J. Environ. Manage.*, 2008, **87**, 600–608.
2. A. C. Edwards and D. Kay, Farmyards an overlooked source of highly contaminated runoff, *J. Environ. Manage.*, 2008, **87**, 551–559; doi:10.1016/j.jenvman.2006.06.027.
3. A. L. Horn, F. J. Rueda, G. Hormann and N. Fohrer, Implementing river water quality modelling issues in meso-scale watershed models for water policy demands – an overview on current concepts, deficits, and future tasks, *Phys. Chem. Earth*, 2004, **29**, 725–737.
4. A. Elshorbagy, R. Teegavarapu and L. Ormsbee, Total maximum daily load (TMDL) approach to surface water quality management: concepts, issues, and applications, *Can. J. Civil Eng.*, 2005, **32**, 442–448.
5. Anon, Council of the European Communities, Directive 2000/60/EC of the European Parliament and of the Council of 23 October 2000 establishing a framework for Community action in the field of water policy, *Off. J. Eur. Union*, 2000, **L327**, 1–72.
6. WHO, *Guidelines for Safe Recreational Water Environments, Vol. 1: Coastal and Freshwaters*, World Health Organisation, Geneva, 2003.
7. G. Rees, K. Pond, D. Kay, J. Bartram and J. Santo Domingo, *Safe Management of Shellfish and Harvest Waters*, International Water Association and WHO, London, 2009, p. 358; ISBN 9781843392255.
8. D. Kay, S. Kershaw, R. Lee, M. D. Wyer, J. Watkins and C. Francis, Results of field investigations into the impact of intermittent sewage discharges on the microbiological quality of wild mussels (*M. edulis*) in a tidal estuary, *Water Res.*, 2008, **42**, 3033–3046; doi:10.1016/j.watres.2008.03.020.
9. D. Kay, A. C. Edwards, B. Ferrier, C. Francis, C. Kay, L. Rushby, J. Watkins, A. T. McDonald, M. Wyer, J. Crowther and J. Wilkinson, Catchment microbial dynamics: the emergence of a research agenda, *Prog. Phys. Geog.*, 2007, **31**(1), 59–76; doi:10.1177/0309133307073882.

10. D. Kay, C. M. Stapleton, M. D. Wyer, A. T. McDonald and J. Crowther, Total Maximum Daily Loads (TMDL). The USEPA approach to managing faecal indicator fluxes to receiving waters: Lessons for UK environmental regulation? in *Agriculture and the Environment VI; Managing Rural Diffuse Pollution, Proceedings of the SAC/SEPA Biennial Conference, Edinburgh*, ed. L. Gairns, C. Crighton and W. A. Jeffrey, 2006, International Water Association, Scottish Agricultural College, Scottish Environmental Protection Agency, Edinburgh, pp. 23–33; ISBN 1901322637.
11. D. Kay, A. T. McDonald, C. M. Stapleton, M. D. Wyer and J. Crowther, Catchment to coastal systems: managing microbial pollutants in bathing and shellfish harvesting waters, in *Catchment Management Handbook*, ed. A. Jenkins and B. Ferrier, 2009, Wiley-Blackwell, Chichester, pp. 181–208; ISBN 9781405171229.
12. D. Kay, R. Lee, M. D. Wyer and C. S. Stapleton, Integrated catchment studies: source identification and modelling, in *Safe Management of Shellfish and Harvest Waters*, ed. G. Rees, K. Pond, D. Kay and S. Domingo, IWA Publishing, London, 2009, p. 358; ISBN 9781843392255.
13. WHO, *Water Safety Plan Manual: Step-by step Risk Management for Drinking-water Suppliers*, World Health Organization, Geneva, 2009; ISBN 9789241562638; http://www.who.int/water_sanitation_health/publication_9789241562638/en/
14. WHO, *Guidelines for Drinking Water Quality, Vol. 2: Health Criteria and Other Supporting Information*, World Health Organisation, Geneva, 4th edn., 2011.
15. G. Rees and D. Kay, Framework for change, in *Safe Management of Shellfish and Harvest Waters*, ed. G. Rees, K. Pond, D. Kay and S. Domingo, International Water Association and WHO, London, 2009, p. 358; ISBN 9781843392255.
16. D. Kay, R. Lee, M. D. Wyer and C. S. Stapleton, Profiling shellfish harvesting waters for public health protection, in *Safe Management of Shellfish and Harvest Waters*, ed. G. Rees, K. Pond, D. Kay and S. Domingo, IWA Publishing, London, 2009, p. 358; ISBN 9781843392255.
17. WHO, *Health Based Monitoring of Recreational Waters: The Feasibility of a New Approach; The "Annapolis Protocol"*, World Health Organisation, Geneva, 1999.
18. C. Ferguson and D. Kay, Transport of microbial pollution in catchment systems, in *Animal Waste, Water Quality and Human Health: WHO - Emerging Issues in Water and Infectious Disease Series*, ed. R. Bos and J. Bartram, International Water Association and WHO, London. (in press).
19. N. Ashbolt and D. Roser, Interpretation and management implications of event and baseflow pathogen data, in *Watershed Management for Water Supply Systems*, ed. M. J. Pfeffer, D. Abs and K. N. Brooks, American Water Resources Association, New York, 2003.
20. P. Cox, M. Griffith, M. Angles, D. Deere, and C. Ferguson, Concentrations of pathogens and indicators in animal faeces in the Sydney watershed, *Appl. Environ. Microbiol.*, 2009, **71**, 5929–5934.

21. C. Ferguson, *Deterministic Model of Microbial Sources, Fate and Transport: A Quantitative Tool for Pathogen Catchment Budgeting,* PhD thesis, University of New South Wales, Australia, 2005.
22. C. Ferguson, N. Altavilla, N. J. Ashbolt and D. A. Deere, Prioritizing – Watershed Pathogen Research, *J. Am. Water Works Assoc.,* 2003, **95**, 92–106.
23. C. Ferguson, A. M. de Roda Husman, N. Altavilla and D. Deere, Fate and transport of surface water pathogens in watersheds, *Crit. Rev. Environ. Sci. Technol.,* 2003, **33**, 299–361.
24. D. Roser and N. Ashbolt, *Source Water Quality and the Management of Pathogens in Surface Catchments and Aquifers,* The Cooperative Research Centre for Water Quality and Treatment Salisbury SA, Australia, 2006.
25. D. Roser, N. Ashbolt, J. Ongerth and G. Vesey, Proficiency testing of *Cryptosporidium* and *Giardia* analyses – an Australian case study, *Water Sci. Technol.: Water Suppl.,* 2002a, **2**, 39–46.
26. D. Roser, J. Skinner, C. LeMaitre, L. Marshall, J. Baldwin, K. Billington, S. Kotz, K. Clarkson and N. Ashbolt, Automated event sampling for microbiological and related analytes in remote sites: a comprehensive system, *Water Sci. Technol.: Water Suppl.,* 2002b, **2**, 123–130.
27. D. J. Roser, N. Ashbolt, G. Ho, K. Mathew, J. Nair, D. Ryken-Rapp and S. Toze, Hydrogen sulphide production tests and the detection of groundwater faecal contamination by septic seepage, *Water Sci. Technol.,* 2005, **51**, 291–300.
28. G. Robinson, R. M. Chalmers, C. M. Stapleton, S. R. Palmer, J. Watkins, C. Francis and D. Kay, A whole water catchment approach to investigating the origin and distribution of *Cryptosporidium* species, *J. Appl. Microbiol.,* 2011, **111**, 717–730.
29. Anon, *Gathering Grounds: Public Access to Gathering Grounds, Afforestation and Agriculture on Gathering Grounds, Report of the Heanage Committee Investigation,* HMSO London, 1984.
30. A. P. Wyn-Jones, A. Carducci, N. Cook, M. D'Agostino, M. Divizia, J. Fleischer, C. Gantzer, A. Gawler, R. Girones, C. Höller, A. de Roda Husman, D. Kay, I. Kozyra, J. López-Pila, M. Muscillo, M. São José Nascimento, G. Papageorgiou, S. Rutjes, J. Sellwood, R. Szewzyk and M. D. Wyer, Surveillance of adenoviruses and noroviruses in European recreational waters, *Water Res.,* 2011, **45**, 1025–1038; doi:10.1016/j.watres.2010.10.015.
31. D. Kay, J. M. Fleisher, R. L. Salmon, F. Jones, M. D. Wyer, A. F. Godfree, Z. Zelenauchjacquotte and R. Shore, Predicting the likelihood of gastroenteritis from sea bathing – Results from randomized exposure, *Lancet,* 1994, **344**, 905–909.
32. D. Kay, J. Bartram, A. Pruss, N. Ashbolt, M. D. Wyer, J. M. Fleisher, L. Fewtrell, A. Rogers and G. Rees, Derivation of numerical values for the World Health Organization guidelines for recreational waters, *Water Res.,* 2004, **38**, 1296–1304.
33. J. M. Fleisher, D. Kay, M. D. Wyer, R. L. Salmon and F. Jones, Non-enteric illnesses associated with bather exposure to marine waters

contaminated with domestic sewage: The results of a series of four intervention follow-up studies, *Am. J. Publ. Health*, 1996, **86**, 1228–1234.

34. L. Fewtrell and D. Kay, *A Guide to the Health Impact Assessment of Sustainable Water Management*, International Water Association, Amsterdam. 2008, p. 320; http://www.iwapublishing.com/template.cfm? name=isbn1843391333

35. L. Fewtrell and D. Kay, An attempt to quantify the health impacts of flooding in the UK using an urban case study, *Publ. Health* 2008, **122**, 446–451; doi:10.1016/j.puhe.2007.09.010.

36. L. Fewtrell, D. Kay, K. Smith, J. Watkins, C. Davies and C. Francis, The microbiology of urban UK floodwaters and a quantitative microbial risk assessment of flooding and gastrointestinal illness, *J. Flood Risk Manage*, 2011, **4**, 77–87; doi:10.1111/j.1753-318X.2011.01092.x.

37. P. R. Hunter, M. Anderle de Sylor, H. L. Risebro, G. L. Nichols, D. Kay, P. Hartemann, Quantitative microbial risk assessment of the risks from very small rural water supplies, *Risk Anal.*, 2011, **31**(2), 228–236; doi:10.1111/j.1539-6924.2010.01499.

38. WHO, *Guidelines for Drinking Water Quality, Vol. 2: Health Criteria and Other Supporting Information*, World Health Organisation, Geneva, 4th edn., 2011.

39. D. Kay, C. M. Stapleton, J. Crowther, M. D. Wyer, L. Fewtrell, A. Edwards, A. T. McDonald, J. Watkins, F. Francis and J. Wilkinson, Faecal indicator organism compliance parameter concentrations in sewage and treated effluents, *Water Res.*, 2008, **42**, 442–454; doi:10.1016/j.watres.2007.07.036.

40. D. Kay, S. Stewart and W. A. Jeffrey, *Evaluation Research into the Effectiveness of Field Best Management Practices at Brighouse Bay*, Final Report for the Scottish Government and SEPA, Scottish Agricultural College and CREH, September 2008.

41. J. Crowther, D. Hampson, I. Bateman, D. Kay, P. Posen, C. Stapleton and M. Wyer, Generic modelling of faecal indicator organism concentrations in the UK, *Water*, 2011, **3**, 682–701; doi:10.3390/w3020682.

42. D. Hampson, J. Crowther, I. Bateman, D. Kay, P. Posen, C. Stapleton, M. Wyer, C. Fezzi, P. Jones and J. Tzanopoulos, Predicting microbial pollution concentrations in UK rivers in response to land use change, *Water Res.* (in press); doi:10.1016/j.watres.2010.07.062.

43. D. Kay, S. Anthony, J. Crowther, B. Chambers, F. Nicholson, D. Chadwick, C. Stapleton and M. Wyer, Microbial water pollution: a screening tool for initial catchment-scale assessment and source apportionment, *Sci. Total Environ.*, 2010, **408**, 5649–5656; doi:10.1016/j.scitotenv.2009.07.033.

44. R. Bos, A. Dufour and J. Bartram, (Eds) *Animal Waste, Water Quality and Human Health:* WHO - Emerging Issues in Water and Infectious Disease series. International Water Association and WHO, London. (in press).

45. D. Kay, J. M. Fleisher, R. L. Salmon, F. Jones, M. D. Wyer, A. F. Godfree, Z. Zelenauchjacquotte and R. Shore, Predicting the likelihood

of gastroenteritis from sea bathing – Results from randomized exposure, *Lancet*, 1994, **344**, 905–909.
46. M. Wyer, D. Kay, J. Watkins, C. Davies, C. Kay, R. Thomas, J. Porter, C. Stapleton and H. Moore, Evaluating short-term changes in recreational water quality during a hydrograph event using combined microbial tracer experiments, environmental microbiology, source tracking and hydrological techniques, *Water Res.*, 2010, **44**, 4783–4795; http://dx.doi.org/10.1016/j.watres.2010.06.047
47. C. M. Stapleton, D. Kay, M. D. Wyer, C. Davies, J. Watkins, C. Kay, A. T. McDonald and J. Crowther, Identification of microbial risk factors in shellfish harvesting waters: the Loch Etive case study, *Aquaculture Res., Special Issue on the Proceedings of the Scottish Aquaculture: A Sustainable Future? Symposium, 21–22 April 2009*, Edinburgh, 2010; ISSN 13652109; doi:10.1111/j.1365-2109.2010.02666.x.
48. J. Crowther, D. Kay and M. D. Wyer, Relationships between microbial water quality and environmental conditions in coastal recreational water: the Fylde coast, UK, *Water Res.*, 2001, **35**(17), 4029–4038; doi:10.1016/S0043-1354(01)00123-3.
49. R. T. Stidson, C. D. Gray and C. D. McPhail, Development and use of modelling techniques for real-time bathing water quality predictions, *Water Environ. J.*, in press; ISSN 1747-6585; doi:10.1111/j.1747-6593.2011.00258.x.
50. C. D. McPhail and R. T. Stidson, Bathing water signage and predictive water quality models in Scotland, *Aquatic Ecosyst. Health Manage*, 2009, **12**(2), 183–186.
51. D. S. Francey and R. Darner, *Procedures for Developing Models to Predict Exceedances of Recreational Water – Quality Standards at Coastal Beaches:* Techniques and Methods 6–B5, United States Geological Survey, Reston, Virginia, 2006, p. 40.
52. Directive 2006/7/EC of The European Parliament and of The Council of 15th February 2006 concerning the management of bathing water quality and repealing Directive 76/160/EEC, *Off. J. Eur. Union*, 2006, **L64**(4.3.2006), 37–51.
53. Centre for Research into Environment and Health, *Source Strengths and Attenuation of Faecal indicators Derived from Livestock Farming Activities: Literature Review*, Report on DEFRA Project WQ0203 Demonstration Test Catchments Initiative, 2010, p. 33; http://randd.defra.gov.uk/Default.aspx?Menu=Menu&Module=More&Location=None&Completed=0&ProjectID=1647052.
54. D. Kay, M. Aitken, J. Crowther, I. Dickson, A. C. Edwards, C. Francis, M. Hopkins, W. Jeffrey, C. Kay, A. T. McDonald, D. McDonald, C. M. Stapleton, J. Watkins, J. Wilkinson and M. Wyer, Reducing fluxes of faecal indicator compliance parameters to bathing waters from diffuse agricultural sources, the Brighouse Bay study, Scotland, *Environ. Pollut.*, 2007, **147**, 139–149; doi:10.1016/j.envpol.2006.08.019.

Pesticides in Modern Agriculture

DAN OSBORN

ABSTRACT

Pesticides have been used since agriculture began and the use or range of metallic and naturally occurring compounds grew slowly and then, in the latter half of the 20th century, rapidly with the advent of synthetic pesticides such as DDT and the cyclodienes. More modern pesticides have been less harmful to the environment as a result of risk-based regulatory regimes and a desire on the part of producers to avoid environmental harm as part of the development of more sustainable framing practices in which pest and diseases are controlled by a range or combination of means. Future challenges from population growth and environmental changes, such as climate change, will mean pesticides will be a necessary part of food production in a future where predicting what pests and diseases of crops need to be dealt with is proving hard to do. Pesticides need to be considered as part of an integrated approach to local solutions that are safe, sustainable, resilient to change and both socially acceptable and economically successful.

1 Introduction

This article looks at how pesticide use originated, during the modern era of agriculture, how it developed and came to be regulated and what principles might govern future pesticide development and management. It provides an overview of the main points that need to be taken into account when thinking about environmental aspects of the use pesticides in crop production, drawing on lessons from current and past uses. It mostly deals with pesticide use since the end of the 1940s when major innovations occurred in the discovery,

Issues in Environmental Science and Technology, 34
Environmental Impacts of Modern Agriculture
Edited by R.E. Hester and R.M. Harrison
© The Royal Society of Chemistry 2012
Published by the Royal Society of Chemistry, www.rsc.org

large-scale manufacture and application of synthetic pesticides. Pesticide use in agriculture is a very complex topic with hundreds of compounds in active use against a global range of pests, weeds and diseases. No single article can do justice to the breadth and depth of the knowledge base and this one is no exception, as it largely excludes discussion of the molluscicides and makes only passing reference to any potential for endocrine disruption.

There is an enormous body of ecotoxicological literature on the effects of pesticides on a whole range of aquatic and terrestrial organisms and this chapter only skims the very surface of those effects of pesticides, covering some of the key environmental factors and events that led to the current regulatory rules on pesticide registration and use. There is also a large amount of literature on the way in which these groups of chemicals become distributed in the environment following original application. It was certainly a very salutary lesson for humanity when residues and metabolites of the early organochlorine pesticides were found in animals at both poles, thousands of miles from the points of application. This observation retains its iconic status in society's view of pesticide impacts on the environment, although there are, after almost fifty years, signs that levels of these substances in Antarctic top predators are now declining if not in that ecosystem overall.[1]

There are many detailed lessons to learn from both ecotoxicological studies and research on the dynamics of pesticide movement and fate, but it is not the aim of this article to review such an extensive literature base. The goal of this chapter is to draw out what we can learn about managing pesticide use in a future in which the pace of environmental and demographic change will require new approaches to the management of crops, natural resources and ecosystem services if enough food is to be produced from limited amounts of land – perhaps as little as 0.3 ha of agricultural land per person by 2050 if the world population exceeds 9 bn. One key to future success will be a need to combine knowledge in an interdisciplinary sense. For example, by combining knowledge from both movement and fate studies and ecotoxicological responses to exposure, it proved possible to determine how important spray drift might be in the environmental effects of pesticides[2,3] and limit damage by giving growers simple advice, improving spraying technology and ending airborne spraying. This illustrates a key to good pesticide management – making good use of interdisciplinary knowledge and combining this with best farming practice. Pesticides were not banned but management practices were adjusted to keep risks to a minimum. Such integrated risk-based approaches are probably a central tenet for the use of pesticides in future.

The use of pesticides in modern high yield agricultural systems is complex and ever-changing as evolutionary forces and human ingenuity compete to keep humanity a step (or two) ahead of a host of other organisms who find the resources supplied for them in modern high yield agricultural systems a considerable convenience. Sophisticated pesticide use is of growing importance. Long gone are the days when farmers and growers might apply pesticides on a prophylactic basis (if such times ever really existed).

Nowadays, in the world's developed economies at least, pesticide applications are often made in response to pest or disease thresholds or trigger levels being passed in ways that – say for fungicides – will minimise the likelihood of resistance developing and, overall, keep costs down by eliminating unnecessary use of pesticides. Pesticides are just one factor involved in producing a good crop. Technological developments aiding pesticide use are also in hand. For example, sensor and information technologies will likely soon be so good that pesticides may only be applied to that part of a field where the crop's health demands it. Such systems can already be used to guide fertiliser applications. Here, technology and good business practices combine to keep levels of use down.

Many different chemical uses are covered by the term "pesticide" and pesticides belong to many different chemical classes. There are at least six different classes of compound in use to tackle fungal diseases alone. If we take farming as a whole, rather than just agriculture, then pesticides are used even to protect animal health – often from insects that are disease vectors. These veterinary uses of pesticides have sets of risks and benefits associated with them that differ markedly from those of pesticide use in crop farming but the two different uses are often, and understandably, treated rather similarly when regulatory authorities are determining the acceptable levels of pesticides that might be found in people's drinking water. Until recently however, veterinary uses have not needed to satisfy all the environmental safety considerations that crop pesticides have had to.

Such complexities mean that all pesticides tend to get lumped together in the public mind, and possibly the political mind, with all the implications that has for the regulation of pesticide use in agriculture – even if some of the impacts of pesticides originate from veterinary uses (*e.g.* the use of organochlorines, organophosphates or pyrethroids to treat sheep and their wool) or from other non-agricultural uses. In terms of overall pesticide use, these other uses are important as sources of concern even if they constitute less than a fifth of the overall use of pesticides worldwide. They include uses in industry, transport systems, local authorities, small businesses and horticulture, as well as in the domestic setting in both gardens and in the home. These uses protect textiles, avoid plant damage to infrastructure systems and deal with pest, disease and nuisance organisms. Some of these uses, such as the use of herbicides in transport and by local authorities, may, because of the speed of run-off from hard surfaces into water courses, have contributed as much or more to the pollution of waterways than has the use of pesticides in agriculture. In some senses, agricultural uses of pesticides have "carried the can" for other non-agricultural uses and, as pressure builds to produce ever more food for a growing world population, it may well be necessary to distinguish much more clearly between effects and costs incurred from agriculture and those that can be attributed to other sectors in which pesticides are used. If we do not do this, then society globally will risk lessening the effectiveness of a vital technological aid that helps maintain humanity's ability to meet the steady demand for ever increasing supplies of food.

2 The Traditional Context of the Agricultural Uses of Pesticides

Pesticides are an unusual group of chemicals because humanity uses them to kill other living things that either: (i) threaten humanity's well-being by eating or destroying the crops we grow to provide food, or (ii) have a role in human disease (such as mosquitoes and flies that are vectors of many human diseases of varying degrees of severity). Few groups of chemicals are released into the environment with the specific aim of killing living things. This is what makes some people think of them in a very negative sense. However, it is worth remembering that many human and veterinary medicines do just that with widely recognised beneficial effects. If one takes a broad view of health and well-being, then pesticides might be considered no less beneficial than some medicines or other groups of pest control agents, such as the biocides (that include the rodenticides used to control rat and mice infestation).

Although people usually think of pesticides as being applied in agriculture, interestingly, some of the first uses of synthetic pest control agents of any kind were to deal with human diseases. One of the first uses of DDT was against the invertebrate vectors that transmitted typhus.[4] The town and population of Naples were sprayed extensively with DDT during the mid-1940s to kill lice that had infested the town after a period of intense military action had destroyed much of the town and its infrastructure. Many lives were saved.

Pesticides also played a significant role in the reduction in malaria in several parts of the world. Again, widespread use of DDT (with its low mammalian toxicity) almost eliminated the diseases from many areas until use was abandoned for a range of economic and environmental reasons (including its effects on certain wild birds) and some concerns – that have proved difficult to quantify – about human health effects following repeated exposure. In 2006, the World Health Organisation[5] recommended the reintroduction of DDT to fight a resurgence of malaria but its use remains controversial partly because it is very persistent in the environment and it bioaccumulates in living organisms and through food webs. This includes its accumulation both in people and breast milk, which many people find to be an unacceptable side effect of the insecticide.

Pesticides are also used to try to reduce impacts of a range of other mosquito and tic borne diseases. Pesticides are often most effective when used as an overall strategy of reducing interactions between people and disease vectors (e.g. through use of mosquito nets treated with insecticides). Here, again, pesticides prove most effective when used in combination with other control measures. The human health, social and economic benefits of using pesticides against nuisance, as well as vector, organisms is substantial as irritation from nuisance bites can be quite debilitating or have important economic consequences, such as reducing tourism to affected places.

In crop protection, pesticides are used in almost all parts of the world, mainly against insects, weeds, nematode worms, molluscs, mites, viruses and fungi.

They are important in much of modern agriculture, particularly in crop systems with high yields. Such systems are rather dependent on effective pesticide use for a number of reasons. This is because pests are "encouraged" to breed, and successfully so, both by the intensity at which the crop is planted and grown, and by the large areas of land carrying the same or similar crops. Such conditions tend to increase the opportunities for insects and other kinds of pests that feed on crops to reproduce readily and reach that range of pest proportions in which yields are reduced. Losses of crop yield due to pests, weeds and diseases (such as fungal rusts in cereals) are thought to represent about 20% of potential yields globally – with much variation regionally and between crops. Losses in some indoor crops may approach the lowest possible figures as the environment in which they grow can be so successfully controlled, especially perhaps when the cropping system uses not just pesticides but other approaches (including biological ones) to control pests. Crop losses of about the same magnitude as in the field occur during storage or transport and here some losses are due to the ravages of pest species (such as rats and mice) that are often controlled through use of biocides *e.g.* rodenticides. Some losses in storage are simply due to poor storage conditions or occur when, say, grains are transported long distances in bulk to markets far from the place the crop grew in. Other losses occur when produce rots before it can get to market or, even, once there.

Historically and currently, pesticides have been very beneficial chemicals in that they make high yield agriculture possible and without such systems for food production it would be impossible to supply the world with enough food. There continues to be debate on this point with proponents of "organic" agriculture arguing that other approaches that depend on crop and soil management can achieve equivalent results. There is no doubt that crops can be grown organically with great success but it is unlikely that such growers would be able to obtain the high yields that deliver relatively low "on-the-plate" costs to the consumer.

Whether pesticides are used in human and animal health or in crop protection there are a number of issues that are common to both types of use. These issues relate to: (i) the adverse and unintended impacts of pesticides on the environment and human health, and (ii) to the problem of resistance, which means that humanity faces a constant battle to produce new chemicals to combat pests both old and new.

Impacts on human health and the environment can arise because of the toxicity spectrum and persistence of pesticides and the tendency of some chemicals, especially those with high octanol–water partition coefficients, to accumulate along food chains or become widely distributed in the environment. Of course, it is quite a challenge to produce a chemical that is uniquely selective to a particular pest species, that persists in the environment only long enough to kill those organisms and that does not accumulate in food chains or contaminate either soil or water. On top of that of course are the problems that arise when the processes of natural selection operate rapidly to produce organisms that are resistant to pesticides.

3   The Changing Nature of Pesticide Use from Earliest Agricultural Times to the Present Day

Pesticides have been in use for many hundreds of years (some uses of sulfur are thought to be nearly as old as agriculture itself which began almost 6000 years ago), with use increasing sharply from the latter half of the 20th century to the present day where active ingredient production is now approaching 3 million tonnes, most of which is used in agriculture.

In the 19th and early 20th century the range of substances employed as pesticides increased. The early compounds used at the very start of the modern era of agriculture included several metals, such as arsenicals, copper compounds (famously used as fungicides against vine diseases) and even mercury and lead, and certain "natural" organic chemical agents, such as nicotine and rotenone (two examples of humanity using nature's own tools to limit pest damage with chemicals that plants had evolved to counter attack insects or other pests). That pesticides were used even in ancient times indicates that even in low intensity systems pesticides may be needed to control damage from pest organisms. Modern organic systems permit the use of some of these chemicals.

In many senses, the modern pesticide era began in the late 1940s with the introduction of DDT to kill the insects that both cause disease in humans and damage crops. DDT was followed by a range of other organochlorine compounds that were found to be lethal to insects. These chemicals included heptachlor, two cyclodienes (aldrin and dieldrin). A range of organometallic compounds, especially compounds of mercury, were introduced as fungicides or sometimes as molluscicides. The environmental impact of organomercurial fungicides varied greatly depending on whether the mercury was linked to an alkyl chain or a phenyl ring. The latter were more likely to have adverse neurological effects and the former were more likely to breakdown in the environment and be excreted or mobilised. The subsequent – an interesting – fate of mercury from such sources is beyond this article's scope.

Pesticide use grew quickly in both Western Europe and North America. Their use in the 1950s and 1960s – together with that of "artificial" fertilisers – made possible a transition to modern, highly mechanised, high yield agriculture. It was during this early period that fears over the adverse effects of pesticides arose.[6] By the mid-1960s international strategies for assessing risk on scientific bases were already being put in place.[7] These recognised the need to investigate and deal with adverse effects but were couched in more scientific terms than Carson had used. This polarisation in approach complicates the consideration of pesticide safety even today.

Benefits were seen from early pesticide use and, at first, no adverse effects were apparent in people or wildlife but over the period from about 1940 to 1960 evidence accumulated that fish and bird kills could be attributed to the use of pesticides. More chronic impacts on the behaviour, embryonic development and reproduction of wildlife became progressively apparent. Research on both sides of the Atlantic began in earnest in the 1960s to quantify effects and it emerged that almost as soon as DDT was introduced into the UK, the eggshell thickness of certain birds, especially predators like sparrowhawk and peregrine falcon, began to decrease. Thin-shelled eggs broke more easily in the nest and so severe was the problem that populations of certain predatory birds began to decline, although for some species the problem was not just pesticides but also persecution. Moreover, the toxic effect was largely due not to DDT itself but to one of its metabolites, *pp'*-DDE. However, DDT/DDE was not the only early pesticide problem faced by birds in general or birds of prey in particular. Reports of acute wildlife poisoning incidents, some involving thousands of birds, were attributed to acute exposure to substances like heptachlor and gradually evidence built up that other organochlorine pesticides, such as the cyclodienes, were able to gradually accumulate in bird tissues and eventually kill them. This meant that some bird species faced a double threat – reproductive success was being reduced by DDT and adult mortality rates were probably being raised due to dieldrin exposure. Fortunately, increasing biochemical, behavioural and ecological evidence from both the laboratory and field were combined to convince both regulatory bodies and producing companies that both cyclodienes and DDT should be either banned or withdrawn from use. Following removal of these compounds from the market in the UK from the mid 1970s to the early 1980s, the affected bird of prey populations recovered. John Sheail[8] sets out the early UK pesticide story very well in his book, and papers by Ratcliffe[9] and Newton and Wyllie[10] recount the decline of these birds and their subsequent recovery following the withdrawal and eventual ban of dieldrin and DDT.

Surprisingly, perhaps, some of the effects of organochlorines, despite their limited solubility in water, were most apparent in aquatic systems. This was due in part to direct overspraying of bodies of water to control mosquitoes, in part to the discharge of industrial waste into bodies of water, as well as to accumulation through food webs which may or may not involve the transport of the pesticide to soil or sediment particles. Tanabe *et al.*[11] illustrate the extent of the ongoing problem for marine systems (although these authors deal with a definition of organochlorines that includes the industrial PCBs as well as pesticides). Xi *et al.*[12] illustrate that is wrong to consider all pesticides that can be classified as "organochlorines" as a single toxicological category and that these compounds can have a range of effects on ecologically relevant physiological and ecological factors. Again, there has been a tendency in the public mind to lump all organic chemicals containing chlorine together and see them as harmful. Despite the complexities involved in understanding the impacts of organochlorine compounds in environmental systems, it was experiences with these chlorinated pesticides (and a few other major classes of chlorinated industrial chemicals, such as the PCBs, which have similar physico–chemical properties) that led to pesticide companies and regulatory bodies implementing environmental safety testing regimes to ensure that pesticides with these properties were unlikely to get to market in any OECD country. Original OECD practice is now widespread, although some middle and high income countries continue to use mixed regimes which draw on both European and

United States experience and which may make their regulatory approaches a little too risk averse.

The properties identified, in broad scientific rather than regulatory terms, were:

  i. a broad spectrum of toxicity;
 ii. persistence for longer than necessary in the environment; and
iii. a log octanol–water partition coefficient > 3, that indicated it would accumulate through food webs such that top predators could suffer toxic effects.

In addition, a special test regime was established for birds to help ensure that no compound was allowed onto the market that had a capacity to induce eggshell thinning, a phenomena whose biochemical basis and target species specificity remained a puzzle until the 1980s and one that may not even yet have been entirely untangled. Khan and Cutkomp[13] and Lundholm[14] set out some of the mechanistic complexity without necessarily taking full account of the interspecies variability that Cooke (1973, 1979)[15,16] points to. The fact is that only some species of bird are susceptible to DDE's eggshell thinning effect. Work on less susceptible species, or those that may even have shown the opposite effect, might not explain the impact on predatory birds. How to protect against such unusual toxic effects remains a theoretical problem even today. Predatory birds are not the easiest to work on in captivity.

Of course, insecticides are not the only kinds of pesticides in use. Weed control by herbicides is as important (if not more so) as pest control. In some parts of the world some weeds can take almost half of a crop, an example being witchweed, a parasite of some crops in Africa and other places, including the United States. Understanding the ecological and biochemical interactions of such weeds is vital if control is to be achieved.[17] Traditional weed control approaches do not work as this plant saps the vigour of crop plants even when below ground, and each plant can produce over a 100 000, and up to 500 000, seeds. New control strategies combining understanding of the nature of plant parasitism with crop management practice may offer the best hope for control of this pernicious weed. The role of herbicides in such instances is limited.

Although many herbicides have low toxicity to mammals and birds, herbicide use has not been without its problems. There have even been human health issues as a result of impurities, such as dioxins, in early types of herbicides. Some of these problems arose from the extensive use in military campaigns and not from the agricultural use of pesticides, for example the fears expressed over the toxicity of 2,3,7,8-tetrachlorodibenzodioxin, an impurity in the mixture of 2,4,5,-T and 2,4-D (Agent Orange) deployed as a forest defoliant during the Vietnam War. Despite the absence of links to agricultural use, such incidents affected the public perception of even relatively non-toxic herbicides. Not all herbicides are non-toxic to humans, however, and deaths have been caused by abuse and misuse of substances such as paraquat. Dioxins do pose health risks but pesticides are not the major source of these, especially as modern production methods prevent such impurities appearing in products.

Herbicides often display opposing chemical properties to those of the old organochlorines in that they are generally more water soluble, however, like organochlorine compounds, they can also persist for some time in the environment. This means that if herbicides get into bodies of water, they can kill plants in these systems to an extent that the integrity of aquatic ecosystems is adversely affected. Removing plants from an ecosystem undermines it completely. This, coupled with the severity of unintended impacts of insecticides, such as dieldrin, on aquatic systems has led a number of regulatory bodies worldwide to set low limits on the levels of pesticides that can be allowed in freshwater systems. These levels are set to protect the most sensitive organisms and, very often, large safety factors are applied that are one or more orders of magnitude below the lowest effect observed in a test species. This is done in environmental protection because only a few test organisms can be used in testing regimes and the aim is to protect the much higher number of species that live in the open environment whose sensitivity to pesticides varies over a similar range to that of the safety factors.

One general theme that emerged from the use of early herbicides is that their widespread use may have led to a decline in the floral biodiversity on farms that in turn may have reduced food supplies for other wildlife. As obvious as this might sound, fully separating these from the effects of insecticides and other groups of pesticides is not easy to achieve because of the interdependencies evident in the food and ecosystem webs of farmland. Several more recent studies illustrate the level of complexity that exists here. For example, Fuller *et al.*[18] point out that the level of biodiversity on organic farms is higher than on conventional farms but that some of this difference may be due to the nature of the locations in which organic farming is conducted (*i.e.* it may not be related to other factors such as lower applications of fertilisers). So establishing simple cause–effect relationships in agricultural systems can be difficult and establishing non-complex facts useful to policy teams or land managers is very challenging. When a large European team[19] looked at a number of factors that might affect biodiversity, pesticides were not picked out as a significant factor. More important factors included nitrogen inputs (*e.g.* from fertiliser use) and the amount of land that was not used for production purposes. This last factor will depend on the nature of the agricultural system being used (*e.g.* fertilised or unfertilised), social and economic choices made by the individual farmer and the terrain and location being farmed. The possibility remains that a very different finding might have emerged had such a study been mounted in the 1950s or early 1960s, for it is sometimes argued that by the time scientific studies on the impacts of pesticides had been begun and suitable studies undertaken, many of the largest declines in biodiversity due to pesticides (including local extinctions of birds of prey) had already occurred and that today's agricultural landscapes lack much of the biodiversity they once had. More positively, it can be argued that today's regulatory system, together with voluntary and evidence-based changes in practice in agriculture and its supply chain, and technological advances in pesticide application means that modern agricultural pesticide regimes are relatively harmless to non-target organisms, especially those that do not make much use of cropped areas or those within a few metres of the field boundary. In many senses this can be seen as a positive result of risk-based regulation.

A major challenge to the grower, apart from pests and weeds, have been crop diseases and these often (but not always) need to be controlled by fungicides (although, for example, insecticides are sometimes used against virus vectors). The major issue with fungicides is not so much human health or environmental effects but the fact that fungicide resistance in target organisms arises all too easily. Resistance occurs when an organism becomes less sensitive to the effects of a given fungicide. Some fungi seem very adept at developing resistance and serious challenges exist for both cereal and rice crops worldwide.

A complex element here is that different varieties of crops have very different susceptibilities to the impacts of pathogens. So fungicides are only one technology humanity can use against fungal pathogens. Plant breeding is another, and both traditional methods and those enhanced by the latest genetic technologies can and will prove useful. The use of genomics in plant breeding does not necessarily mean that the plant varieties that result are genetically engineered; genomic technologies can simply be used to speed up the progress of favourable trait selection in new plant varieties. The situation is changing rapidly both globally and in individual countries, such as in the UK where new guidance for growers was issued in 2011 (http://www.pesticides.gov.uk/ Resources/CRD/Migrated-Resources/Documents/F/FRAG_Cereals_Resistance_ guidelines_v4_March_2011.pdf).

Problems with resistance to fungicides mean that farmers and growers are now advised on strategies for minimising the risk of resistance developing. Approaches include: using mixtures of compounds, and making sure applications occur at the right time and that the correct doses are used. When mixtures are used, fungicides with different modes of action are normally included.

Somewhat ironically, as a result of the many different modes of fungicide use there is the potential for other non-target organism effects to emerge due to the use of mixtures of compounds. For example, it has been shown that some members of a common group of fungicides – the conazoles – can interfere with the ability of birds to metabolise organophosphate insecticides. Johnston et al.[20,21] showed that, under controlled conditions, pre-treatment with prochloraz increased the toxicity of a subsequent exposure to the insecticide, malathion. Although bird mortality from exposure to organophosphorous insecticides can occur in the field (e.g. the deaths of hundreds of an unexpectedly susceptible species of geese in the UK),[22] it seems unlikely that interactive effects such as those found in controlled conditions are likely to occur in the field. However, the possibility of such interactive effects is an active research field that has been used to probe the underlying toxicological response patterns of a range of organisms.

4   Risks to Human Health from Pesticide Use in Agriculture

To protect human health from pesticide impacts, almost every regulatory authority sets limits on pesticide concentrations in drinking water and food items. An important facet of this limit setting exercise is that safety factors of several orders of magnitude are applied to the toxicity data derived from a range of species that are thought to be suitable as test models for humans. Once again, the safety factors can take acceptable pesticide levels in food and water far below the concentrations likely to do harm. Human health effects arising from low-level exposures have proved difficult to establish unequivocally, although association with some neurological conditions seems likely as a result of exposure to pesticide types some of which, at least, are now withdraw from use. Dick[23] indicates how complex it can be to establish clear cause and effect relationships on such matters, in this case with respect to Parkinson's disease.

Hamilton *et al.*[24] illustrate the complexity of setting such limits for pesticides to protect human health in a global context and some of the differences between regulatory regimes in different parts of the world. The complexity of the pesticide regulatory regimes is far too complex a subject to enter into here at any level other than to cover their main guiding principles. Setting acceptable limits in food is just as complex, and details can be found in the *Codex Alimentarius* (http://www.codexalimentarius.net/web/index_en.jsp) with sections on pesticide residues dealt with jointly by the World Health Organization (WHO) and the Food and Agriculture Organization of the United Nations (FAO).

Perhaps the most important impact of the growing body of scientific knowledge on the agricultural use of pesticides is the emergence of a risk-based approach to pesticide management and the protection of human health and the environment. The importance of a risk-based management of pesticides cannot be stressed enough, as such systems are based on the evidence of effects on organisms. Regulatory regimes based on other approaches, such as those that currently operate within the European Union, have an inherent tendency to be based on the potential of groups of chemicals to present a hazard, rather than a risk, to the environment or human health. The importance of establishing cause-and-effect and source-receptor pathways are minimised in such approaches, and there is an emphasis towards blanket prohibition of groups of substances on the basis that a chemical might cause harm. All motor cars pose a hazard to the environment and human health, but we do not consider banning them. Instead we adopt a risk management approach that balances risks and benefits. This is the best way forward for management of all forms of technology. Of course such approaches allow society to decide that the risk or cost of using a particular technology are too great and in such instances the technology – be it a chemical, a machine or a process – can be banned.

Moreover, in some hazard-based regulatory regimes, inappropriate politicisation of the evidence base, and an application of the precautionary principle far different to that for which it was originally developed, can turn the principles science used to develop robust risk-based approaches almost on their head. Whereas, in risk-based approaches properties, such as the log octanol–water partition coefficient, were used as guides to risk-based decision making, in hazard-based systems such data become more absolute determinants of the decision making outcome. The practical result of hazard-based regimes means that if there is only a possibility that a compound has one of a certain set of physico–chemical or toxicological properties, there is a regulatory predisposition against its use even if there is little chance of risk from substances being realised. As applied today, hazard-based systems tend to drive whole groups of compounds off the market, potentially leaving growers – and therefore consumers – exposed to new risks, such as those resulting from food shortages or higher prices or even to food contaminated with fungal toxins.

However, dealing with hazard-based regulatory regimes is only one of the challenges that sustainable pesticide use will need to account for, although the agro-chemical sector will not be alone in meeting the challenges that lie ahead from environmental and demographic changes in the middle and later 21st centuries.

5   Pesticide Use in Current Agricultural Systems – A Changing and Challenging Context

Towards the end of the 20th century, pesticide use took a new direction when genetic material for herbicide resistance was moved between different species to enable weed control agents to be used in crops that would normally succumb to the effects of the herbicide. These genetically modified crops caused enormous public controversy in Europe and were banned on a range of grounds. Scientific field trials had suggested environmental harm might result from the way such crops were to be managed.[25–27] Here the perceived risk in the public mind was from both the genetic modification and from the extension in the use of pesticides into new areas. This can be seen as a special case of dealing with risks arising from two sources and, more or less at the same time, ecotoxicologists, and toxicologists more generally, had started to develop modern approaches to examining both how exposure to mixtures of chemicals could be approached and how toxicological evidence could be combined with population biology to produce knowledge that would help in time develop more realistic risk assessments for the environment. Pharmacologists had developed theories of interactions between substances in the late 1930s.[28] This growing interest in how organisms and even whole systems might respond to two or more pressures applied simultaneously, and what this might mean for organisms when exposed under field conditions, stimulated studies into not only mixtures of chemicals[29] but also on how ecotoxicological responses (such as biomarkers) could be used to predict ecologically significant effects on the life histories of organisms.[30–33] These studies have now been extended to pesticide mixtures, if only in model organisms[34] and with test soils from agricultural sites.[35] Such work holds out some prospect of developing test regimes that are both ecological relevant and that might, perhaps, provide more exact estimates of risks to populations of wild organisms. They use a range of techniques from feeding studies on soil invertebrates (*via* bait lamina systems) to genomic approaches. None of these techniques have yet been applied extensively in regulatory pesticide testing but they may be very helpful in future. In time, such approaches may reduce the need for the large safety factors that currently need to be applied to pesticide regulation. These approaches may also increase the predictive power of test regimes and, combined with knowledge of chemical structures, reveal useful information about what types of compounds are least likely to cause problems to non-target organisms and most problems for pests. If, after much further research, these approaches do succeed, this will in part be due to their strong theoretical and mechanistic basis. Many early problems with pesticides probably arose because knowledge had a more empirical base and it was, therefore, much more likely that surprising results would be uncovered or come to light.

Such studies on soil organisms and other invertebrate groups may also have practical results for pesticide regulation in the near future. Much more attention is already paid to effects on earthworms and certain pollinating insects in modern regulatory systems than was the case a few years ago, although it still remains difficult to interpret the ecological significance of both lethal and sub-lethal test results for wild populations despite the interest in such topics extending back at least 35 years.[36] This is because there is still only limited knowledge about the dynamics of wild populations and how, at landscape scales, these respond to, say, a pesticide application killing non-target organisms in a cropped area. Such losses could be replaced in the course of a season (or in the next year) by either reproduction or immigration from surrounding land. This is where sub-lethal measures can be important because, if there are no or low-level ecologically important (*e.g.* life history) effects, then it should be possible to permit use. The problem comes when sub-lethal measures in the mid- to high-range are found. This is where models of the kind developed by Kammenga *et al.*[32] and Jonker *et al.*[29] could find wider application, although the scope of the issues with pesticides is still being drawn out for groups of organisms such as bees[37] where determining routes of exposure, as well as meaningful end-points, are still problematic.

6 Future Pesticide Use and Approaches to their Regulation and Management

Pesticide use in agriculture has thus moved from being considered first as an absolute boon, to becoming a problem involving non-target organisms and, to a more limited extent, human health. Although the modern regulatory system protects both the environment and human health, questions remain over the way in which risk assessment has to be conducted with semi-political (with a small 'p') arguments deployed on all sides to deal with some of the issues that as yet science has found intractable. Despite all the efforts to make pesticide risk assessment more realistic and ecologically relevant, for instance, by the use of probabilistic modelling approaches such as those of Hart and colleagues,[38] pesticide use will face increasing rather than decreasing challenges if it is dealt with in isolation from a number of other forces affecting modern and future agriculture. This is because pesticides are of only part of the answer to the challenge of feeding the 9 or even 10 billion people expected to inhabit the planet by the middle of this century and, likewise, only part of the answer for controlling disease. Even today, many hundreds of millions of people struggle to get enough to eat. Some are on starvation rations. Such situations are affected by many environmental, social, governance and economic factors, and there is a real challenge in dealing with two weird anomalies: (i) that there are as many obese people as there are those close to starvation, and (ii) that in some parts of the world food is thrown away and in others there is patently a failure of supply chains even in times when overall global food production is high. Pesticides are part of this much larger mix, a major force in enabling food production, and must be seen and managed as such.

In addition to the problems caused by the growing world population (that in itself suggests that food shortages are not limiting human population growth) and the demographic pressures this is leading to, the environment in which agriculture is carried out, and on which its success ultimately depends (all crops fail without water), is changing too – possibly at an accelerating rate. Almost certainly, the overall rate of change is greater than previously experienced by humanity itself or its agricultural systems. Modern civilisations and their agriculture have developed in a period of relative climatic stability. The current and upcoming change is caused by increased carbon dioxide levels in the atmosphere steadily changing the heat balance of the planet such that surface temperatures are now headed for at least two degrees above the pre-industrial average. This is very much an average and the figure masks many diurnal, seasonal and regional differences. Temperature changes in polar regions may exceed six degrees. These temperature changes are already affecting biodiversity and agriculture, with growing seasons in Europe extending by as much as 10 days over the past few decades. Effects are not the same on all trophic levels, and there are already examples of vectors' distributional ranges changing in response to warming or of them becoming more effective as disease transmission agents.

This combination of environmental and demographic challenges, including the increasing (coastal) urbanisation of the planet's population, will pose substantial challenges for agriculture and the whole food supply chain. What combination of temperature and rainfall patterns will prevail at any point on the globe in future is proving difficult to predict very far in advance. But pesticide companies need foresight of such changes – as does the rest of agriculture – in order to plan what varieties of crop will grow where, what yields are likely to be and what pests and diseases might need to be combated. It will also be necessary to develop high yielding agricultural systems that do not harm ecosystems, and to supply ecosystems services[39] that may best be delivered by developing a multifunctional approach to landscapes with appropriate degrees of "land-sharing" and "land sparing".[40]

In short, future food production systems will need to be:

   i. safe for workers, consumers and the public;
  ii. sustainable in terms of its use of natural resources;
 iii. resilient to changing climatic and demographic conditions; and
  iv. appropriate to local economic and social conditions.

This is a very tall order for food production systems to meet. Unless such issues are addressed, however, future agriculture may well fail to produce enough food because, for various reasons, soils will have been destroyed or eroded and thresholds for sustainability exceeded, not least because of pressures arising from a sharp decline in the amount of agricultural land per head of population. By 2050, this may be as little as 0.3 hectares per person, and by that time the suitability of large areas of the world for agriculture will have been adversely affected by reduced rainfall.

What is clear from current dialogues about the future of food production and agriculture in a rapidly changing world is that new approaches to food production are needed. Traditionally, the main ecosystem service delivered by agriculture – the provisioning service of the production of crops that are turned into foods – has been delivered successfully by modifying a number of the supporting services, like the biogeochemical cycles for nitrogen and phosphate, so that crops can be fertilised and produce higher yields. Even in present day agriculture, there is some evidence that sustainable thresholds for pollution in freshwaters and ground waters resulting from the use of nitrogen and phosphate fertilisers have been exceeded. Pesticides themselves modify regulating ecosystem services by reducing crop losses from pests. Furthermore, modern agricultural systems were first developed with a pure focus on production that continued in Europe until modifications to the Common Agriculture Policy were made that enabled payments to be made to farmers in return for, in effect, the delivery of ecosystem services relating to biodiversity. These changes to the regulatory framework of agriculture, as helpful as they are, still in many senses leave considerations of environmental resources and systems as one of the last things to be considered. This is often the case with advanced economic systems. For too long, because technological developments have been designed at meeting human needs and have not considered planetary resources as an integral part of the economic system, environmental issues have come to be seen as a problem rather than as part of the an integrated solution.

In summary, risk-based approaches to pesticide management are likely to be the most useful in that they provide a greater opportunity for promoting innovative solutions to dealing with the basic challenges society faces, such as the provision of shelter, food and water. Hazard-based systems tend to be too limiting in a market-based economic system and might even be so in some other economic systems where time and resources are unlimited. Risk-based systems that encourage innovation are much more likely to be able to deal with the basic challenges that are likely to become more of an issue as the world population grows, its demography and economic and social needs change, and as the environment itself changes as a result of the pressures placed upon it by the growing population and its continuing reliance on fossil fuels or other resource-intensive forms of producing energy. In addition, the world is now a highly interconnected and interdependent place that has become used to working in a way that delivers goods and services just when needed ("just in time") and, as a result, failures – say in crop yields – in one part of the world can no longer be thought of as an isolated incident affecting faraway places.

The lack of global or even regional grain storage systems means a crop failure in one place soon affects food prices in the rest of the world. There have been several examples of this kind of issue in recent years, the latest of which were droughts in central Russia that stopped exports of Russian grain to other places and caused global food price rises in goods using cereals. We need to do everything we can to avoid major crop failures and yield losses, and the wise use of pesticides is one tool we have at our disposal.

Future directions in pesticide use must consider pesticides as part of an overall crop and food system. Food needs to be produced in sufficient quantity and must be safe for human consumption, sustainably produced, resilient to future environmental challenges and produced in a way that is appropriate to the local environmental, as well as the social and economic conditions of the producers. It will not be possible to have solutions that apply to the whole world. Such integrated solutions will be challenging to produce and find and will involve working in partnership with producers, consumers and regulatory authorities. Research-led dialogues about pesticide use will have an important part to play in future pesticide regulatory regimes.

References

1. N. W. van den Brink, M. J. Riddle, M. van den Huevel-Greve and J. A. Van Franeker, *Mar. Pollut. Bull.*, 2011, **62**, 128–132.
2. B. N. K. Davis, M. J. Brown, A. J. Frost, T. J. Yates and R. A. Plant, *Ecotoxicol. Environ. Saf.*, 1994, **27**, 281–293.
3. R. H. Marrs, C. T. Williams, A. J. Frost and R. A. Plant, *Environ. Pollut.*, 1989, **59**, 71–86.
4. C. M. Wheeler, *Am. J. Public Health*, 1946, **36**, 119–129.
5. World Health Organization, *WHO gives Indoor Use of DDT a Clean Bill of Health for Controlling Malaria*, 2006; http://www.who.int/mediacentre/news/releases/2006/pr50/en
6. R. Carson, *Silent Spring*, Houghton Mifflin, New York, 25th Anniversary edn, 1962.
7. N. W. Moore, *J. Appl. Ecol.*, 1966, **3**, Supplement.
8. J. Sheail, Pesticides and nature conservation: the British experience 1950–1957, *Monographs on Science, Technology and Society*, Clarendon, Oxford, 1985, No. 4, p. 276.
9. D. A. Ratcliffe, *J. Appl. Ecol.*, 1970, **7**, 67–115.
10. I. Newton and I. Wyllie, *J. Appl. Ecol.*, 1992, **29**, 476–484.
11. S. Tanabe, H. Iwata and R. Tatsukawa, *Sci. Total Environ.*, 1994, **154**, 163–177.
12. Y.-L. Xi, Z.-X. Chu and X.-P. Xu, *Environ. Toxicol. Chem.*, 2007, **26**, 1695–1699.
13. M. H. Khan and L. K. Cutkomp, *Arch. Environ. Contam. Toxicol.*, 1982, **11**, 627–633.
14. C. E. Lundholm, *Comp. Biochem. Physiol., Part C: Pharmacol., Toxicol. Endocrinol.*, 1997, **118**, 113–128.

15. A. S. Cooke, *Environ. Pollut.*, 1973, **4**, 85–152.
16. A. S. Cooke, *Environ. Pollut.*, 1979, **19**, 47–65.
17. G. R. Stewart and M. C. Press, *Annu. Rev. Plant Physiol. Plant Mol. Biol.*, 1990, **41**, 127–151.
18. R. J. Fuller, L. R. Norton, R. E. Feber, P. J. Johnson, D. E. Chamberlain, A. C. Joys, F. Mathews, R. C. Stuart, M. C. Townsend, W. J. Manley, M. S. Wolfe, D. W. Macdonald and L. G. Firbank, *Biol. Lett.*, 2005, **1**, 431–434.
19. R. Billeter, J. Liira, D. Bailey, R. Bugter, P. Arens, I. Augenstein, S. Aviron, J. Baudry, R. Bukacek, F. Burel, M. Cerny, G. De Blust, R. De Cock, T. Diekötter, H. Dietz, J. Dirksen, C. Dormann, W. Durka, M. Frenzel, R. Hamersky, F. Hendrickx, F. Herzog, S. Klotz, B. Koolstra, A. Lausch, D. Le Coeur, J. P. Maelfait, P. Opdam, M. Roubalova, A. Schermann, N. Schermann, T. Schmidt, O. Schweiger, M. J. M. Smulders, M. Speelmans, P. Simova, J. Verboom, W. K. R. E. Van Wingerden, M. Zobel and P. J. Edwards, *J. Appl. Ecol.*, 2008, **45**, 141–150.
20. G. Johnston, G. Collett, C. Walker, A. Dawson, I. Boyd and D. Osborn, *Pestic. Biochem. Physiol.*, 1989, **35**, 107–118.
21. G. Johnston, C. H. Walker, A. Dawson and A. Furnell, *Functional Ecol.*, 1990, **4**, 309–314.
22. P. I. Stanley and P. J. Bunyan, *Proc. R. Soc. London, Ser. B*, 1979, **205**, 31–45.
23. F. D. Dick, *Br. Med. Bull.*, 2006, **79–80**, 219–231.
24. D. J. Hamilton, Á. Ambrus, R. M. Dieterle, A. S. Felsot, C. A. Harris, P. T. Holland, A. Katayama, N. Kurihara, J. Linders, J. Unsworth and S.-S. Wong, *Pure Appl. Chem.*, 2003, **75**, 1123–1155.
25. L. G. Firbank, M. S. Heard, I. P. Woiwod, C. Hawes, A. J. Haughton, G. T. Champion, R. J. Scott, M. O. Hill, A. M. Dewar, G. R. Squire, M. J. May, D. R. Brooks, D. A. Bohan, R. E. Daniels, J. L. Osborne, D. B. Roy, H. I. J. Black, P. Rothery and J. N. Perry, *J. Appl. Ecol.*, 2003, **40**, 2–16.
26. M. S. Heard, C. Hawes, G. T. Champion, S. J. Clark, L. G. Firbank, A. J. Haughton, A. M. Parish, J. N. Perry, P. Rothery, D. B. Roy, R. J. Scott, M. P. Skellern, G. R. Squire and M. O. Hill, *Philos. Trans. R. Soc. London, Ser. B*, 2003, **358**, 1833–1846.
27. C. Hawes, A. J. Haughton, J. L. Osborne, D. B. Roy, S. J. Clark, J. N. Perry, P. Rothery, D. A. Bohan, D. R. Brooks, G. T. Champion, A. M. Dewar, M. S. Heard, I. P. Woiwod, R. E. Daniels, M. W. Young, A. M. Parish, R. J. Scott, L. G. Firbank and G. R. Squire, *Philos. Trans. R. Soc. London, Ser. B*, 2003, **358**, 1899–1913.
28. C. I. Bliss, *Ann. Appl. Biol.*, 1939, **26**, 585–615.
29. M. J. Jonker, C. Svendsen, J. J. Bedaux, M. Bongers and J. E. Kammenga, *Environ. Toxicol. Chem.*, 2005, **24**, 2701–2713.
30. D. J. Spurgeon and S. J. Hopkin, *Ecotoxicol. Environ. Saf.*, 1996, **35**, 86–95.
31. D. J. Spurgeon and S. P. Hopkin, *Ecotoxicology*, 1999, **8**, 133–141.
32. J. E. Kammenga, D. J. Spurgeon, C. Svendsen and J. M. Weeks, *Oikos*, 2003, **100**, 89–95.

33. D. J. Spurgeon, C. Svendsen, J. M. Weeks, P. K. Hankard, H. E. Stubberud and J. E. Kammenga, *Environ. Toxicol. Chem.*, 2003, **22**, 1465–1472.
34. C. Svendsen, P. Siang, L. J. Lister, A. Rice and D. J. Spurgeon, *Environ. Toxicol. Chem.*, 2010, **29**, 1182–1191.
35. M. J. Santos, R. Morgado, N. G. Ferreira, A. M. Soares and S. Loureiro, *Ecotoxicol. Environ. Saf.*, 2011, **74**, 1994–2001.
36. C. A. Johansen, *Annu. Rev. Entomol.*, 1977, **22**, 177–192.
37. H. M. Thompson, *Pest Manage. Sci.*, 2010, **66**, 1157–1162.
38. A. Hart, *Probabilistic Risk Assessment for Pesticides in Europe: Implementation & Research Needs*, Report of the European workshop on Probabilistic Risk Assessment for the Environmental Impacts of Plant Protection Products (EUPRA), The Netherlands, June 2001, Central Science Laboratory, York, 2001, p. 109.
39. UK NEA, *The UK National Ecosystem Assessment: Synthesis of the Key Findings*, UNEP-WCMC, Cambridge, 2011, p. 86.
40. J. A. Hodgson, W. E. Kunin, C. D. Thomas, T. G. Benton and D. Gabriel, *Ecol. Letts*, 2010, **13**, 1358–1367.

Balancing the Environmental Consequences of Agriculture with the Need for Food Security

IAN CRUTE

ABSTRACT

The UK Government's *Foresight* Report on *The Future of Food and Farming* stressed that environmental sustainability was an essential component of any changes in the global food system directed towards the pressing requirement to scale-up levels of productivity significantly. This paper examines in a historical and contemporary context what some of the issues are in attempting to achieve a balance between producing more food and the fact that agricultural systems have environmental impacts, and always have had. Impacts on biodiversity, water and climate change through the emissions of greenhouse gases are the primary issues confronted. A primarily UK-centric approach is taken to exploring the concept of "Sustainable Intensification" as it might assist in addressing biodiversity conservation and the need to mitigate global climate change by reducing net greenhouse gas emissions from agricultural systems. The need for more data and less dogma in addressing the complex of issues associated with the environmental impact of food production systems is stressed, as is the need to adopt a systems-based approach to land use which seeks to identify, quantify and contextualise the extent to which trade-offs are required. In seeking to establish a better carbon balance from land-based activities, an alternative is suggested to the rather unimaginative and one dimensional approach of setting targets for reductions in greenhouse gas emissions.

Issues in Environmental Science and Technology, 34
Environmental Impacts of Modern Agriculture
Edited by R.E. Hester and R.M. Harrison
© The Royal Society of Chemistry 2012
Published by the Royal Society of Chemistry, www.rsc.org

1 Preamble

1.1   The Need for Food Security

Much has recently been written and spoken about the urgent need to elevate global food production significantly because of both population growth and the dietary change that is a consequence of rising standards of living in developing countries. At the same time, the need to respond to and manage the increasing demand for fresh water and energy while attempting to mitigate and adapt to the threat of climate change presents an enormous set of challenges, which have been thoroughly explored in the recently published UK Government's *Foresight*[1] report on the *Future of Food and Farming*. If the upward trajectory of global development (and associated decline in birth rate) is maintained while simultaneously ensuring that food production and distribution keeps pace with demand, the prize of a stable (perhaps even declining) and nutritionally secure global population is a prize well worth striving for; although not in itself sufficient, food security is an absolutely necessary requirement if future generations are likely to enjoy political stability and freedom from conflict.

1.2   The Importance of Environmental Sustainability and a Role for the UK

Foremost in the thinking about how humanity should act to meet these challenges is the need to recognise the importance of environmental sustainability and particularly ensuring the integrity of the ecosystem services on which humanity depends. With the publication of the *UK National Ecosystem Assessment*[2] following so closely after the *Foresight* report, these issues have been well exposed to scrutiny and debate in the UK. This level of engagement has meant that an active and healthy dialogue has rapidly developed between those individuals and organisations concerned with a whole spectrum of issues that span international development, food security, national agricultural competitiveness and food production, conservation of biodiversity, water quality and availability, bioenergy production, climate change and sustainability in general. The UK has much to offer internationally in both practical and intellectual terms to the way policies that address the future can be developed and implemented, despite the fact that the country is occupied by less than 1% of the global population and occupies less than 1% of the world's agricultural land (Table 1).

Although a relatively small land mass, the UK is densely populated and highly urbanised with enormous geodiversity[3] on which is founded diverse land use including a wide variety of agricultural systems. This includes a relatively high proportion of improved pasture compared to other parts of western and northern Europe, large areas of semi-natural upland grassland, productive arable land, but a relatively low proportion of woodland and almost no unmanaged natural terrestrial ecosystems (Table 1). It is interesting

Table 1 Comparative human population, land availability and usage in UK, N+W Europe, China and the World. [Data from FAOStat, 2011 (year of access – not data)].

	UK	N+W Europe	China	World
Population (millions)	62 (0.9% of world)	291 (4.2% of world)	1340 (19.2% of world)	6979
Land (millions of hectares)				
Total Land Area	24 (0.2% of world)	273 (2.1% of world)	932 (7.2% of world)	13003
Agricultural land	17 (0.3% of world)	93 (1.9% of world)	524 (10.7% of world)	4889
Forest (+Savannah)	3 (12.5% of area)	105 (38.5% of area)	204 (21.9% of area)	4039
Crop land	6 (25.0% of area)	55 (20.1% of area)	124 (13.3% of area)	1533
Pasture land	11 (45.8% of area)	38 (13.9% of area)	400 (42.9% of area)	3356
Other	4 (16.7% of area)	75 (27.5% of area)	204 (21.9% of area)	4088
Land per person (hectares)				
Crop land	0.097 (56% less than world)	0.189 (14% less than world)	0.092 (58% less than world)	0.220
Pasture	0.177 (63% less than world)	0.131 (73% less than world)	0.298 (38% less than world)	0.481

to note that despite the huge differences in land area and population, the UK shares with China a low proportion of agricultural land per capita (compared to the global average) which is particularly evident for crop land. This illustrates the relevance of endeavours in the UK to balancing the environmental consequences of agriculture with the need for food security; the subject of this paper.

The way land is used and managed sits at the very foundation of providing food security while ensuring continuing ecosystem functionality; as a small but well-resourced country, the UK has much to contribute to finding the route through which analysis of the complex set of interactions that impinge upon this issue may lead to a satisfactory resolution. It is to be hoped that such an endeavour will have impact beyond the confines of one small country. Looking at the future of the UK, highly productive and resource-use efficient agricultural systems will be necessary. These will need to be constructed and managed in such a way as to deliver the food requirements of a population that could reach 70 million while maintaining or restoring environmental integrity to ensure continuing delivery of essential ecosystem services (that include the provisioning of food and cultural services that enrich life). This paper seeks to use the example of the UK to draw out thinking about the way in which it should be possible to gather the knowledge and develop the principles for approaches that will be necessary to arrive at a well-founded trajectory for national and international action.

2   Agriculture's Environmental Impact and a Summary of the Issues

2.1   Some Terminology

For the purpose of this paper, agriculture is used as an inclusive term to encompass the production of domesticated livestock as well as all edible and non-edible crops grown at scales that exceed areas generally described as "gardens" or "allotments". The term excludes aquaculture and forestry (trees grown for timber) which are of course important but beyond the scope of this paper. For clarity, "production" refers to gross output and does not imply anything about efficiency; "productivity" is a measure of efficiency and used to refer to the unit of production (*e.g.* tonne of wheat) per unit of input (such as land or labour) or environmental cost (including waste or emissions from the system). Hence, for example, litres of milk per cow, or per hectare, or per Kg CO_2 equivalents are all different measures of efficiency and productivity.

2.2   Man-managed and Natural Ecosystems – Competition for Photosynthate

In essence, agriculture is founded on the domestication of plant and animal species that commenced approximately 12 000 years ago[4,5] and is a man-managed system, the primary function of which is to convert solar energy into chemical energy (in various forms) for utilisation by mankind as foodstuff and other products including fibre and fuel. Ruminant and monogastric livestock species represent important intermediates in this energy conversion process because of their ability to feed on forms of plant or animal biomass that are inedible by humans. A variety of domesticated animal species are therefore a source of valued human food and other co-products both before (*e.g.* wool, eggs and milk) and after slaughter (meat and hides). It is also worthy of note that large domesticated animals are also an important source of power in the agricultural production systems of less developed countries.

From the very earliest of times, cultivation of crops and the keeping of livestock have impacted on natural ecosystems; forests were cleared by burning; grasslands were grazed by domesticated herds and flocks; soil was tilled to enable crop cultivation; and water was diverted from rivers and lakes for irrigation. Over the millennia, human nutrition globally became increasingly dependent on agriculture as it substantially substituted for hunting wild animals and gathering wild plants. In common with other animal species, the growth of the human population was primarily constrained by the availability of adequate nutrition (as well as disease when local population densities increased in settlements). Agriculture relaxed the nutritional constraint on population size and enabled acceleration in the processes leading to establishment of settlements and human civilisations in an urban setting.

Historically, there has been an intimate connectivity between the rise and fall of civilisations and their capacity to produce, reliably and predictably, sufficient food to sustain growing populations over successive generations;[6,7] the latter being a highly pragmatic description of sustainable agriculture. Unsustainable systems of food production (and other biomass such as fuel and timber for construction) that are based on the over-exploitation of natural resources leading to environmental degradation have been at the root of human migration and conflict for millennia. This situation continues to the present and can be exacerbated in countries with expanding populations where feed-back effects on water availability and soil fertility may mean that conditions already unfavourable to highly productive agriculture become progressively worse.

The environmental consequence of agriculture in a historical and modern-day setting has its roots in the competition for photosynthate between the human species and other species through the appropriation of plant biomass or net primary productivity (NPP). The more biomass that humans utilise or destroy or the more they adversely impact the natural global capacity for biomass productivity, the less energy is available to support those other forms of life that ultimately also depend on solar energy conversion. This competition is essentially manifest in the way in which mankind chooses to use land (*e.g.* for crops, forests, livestock, cities *etc.*) and particularly the way this land is actually managed to meet the purpose to which it is put. Furthermore, over the millennia, man has genetically improved the domesticated species of crops and livestock to be increasingly productive and well-adapted to the agricultural environments that have been specifically created for them. Mankind has also ensured, for his own benefit, the reproductive success of the relatively few species on which he has come to depend and which, in terms of individual plants or animals, are now amongst the most numerous on the planet. It is estimated that less than 25% of the land area of the earth is now "wild" (*i.e.* unmanaged in some way by man) and that this accounts for just 11% of NPP.[8] It is also estimated that humans now appropriate almost 25% of global NPP.[9]

The mutual high level of dependency of the vast majority of humans on a small number of specialised plants and animals can be viewed as a form of well-tuned symbiosis. However, this cannot be divorced from the man-managed agro-ecosystems which necessarily require a myriad of other organisms to enable them to function by, for example: regulation of pest, parasite and pathogen populations, pollination of crops, chemical detoxification and cycling of essential plant nutrients (collectively referred to as "ecosystem services").

2.3   The Application of Science – Manipulating Genotype and Environment

Over time, mankind has been able to exploit with ever increasing efficiency, the interaction between crop or livestock genotypes and the man-modified environment by purposefully tuning it both to his own advantage and, often, to the disadvantage of species that share the same space. In the last 150 years, the application of science has massively accelerated the process by which the genetic potential of domesticated species can be increased through selective breeding and then realised in intensively managed agricultural systems (exemplified by the "Green Revolution"). It can be argued that this has contributed significantly to a diminution of the potential adverse impact on the environment that might otherwise have resulted had larger land utilisation been required to meet the human demand for food.[10] As the data in Table 2 shows, it is a remarkable achievement that cereal yields have kept ahead of population growth and that the increase in land devoted to production of these staple crops has elevated by just 8% in half a century.

Table 2   Changes in global population, cereal areas, yields and grain per capita (1960–2010). (Data from FAOStat, 2011).

Year	1960	1970	1980	1990	2000	2010	50 year growth (%)
Population (billions)	2.97	3.70	4.40	5.10	6.10	6.80	
% increase		25	19	16	20	11	129
Area (m hectares)	647	675	717	708	673	699	
% increase		+4	+6	−1	−5	+4	8
Yield (tonnes per hectare)	1.35	1.77	2.16	2.76	3.06	3.57	
% increase		31	22	27	11	16	164
Quantity per person per annum (Kg)	294	323	352	383	338	367	
% increase		+10	+9	+9	−11	+9	+25

However, the statements above tend to ignore the fact that elevated livestock productivity coupled with increased crop productivity have allowed a major change in the human diet to occur in some parts of the industrialised world and increasingly in those countries now on a rapid development trajectory. Increased consumption of meat, eggs and dairy products has been made possible by increased forage crop productivity and the availability of refined animal feed in the form of cereals and pulses additional to that used directly for human consumption. The environmental consequences of the increased numbers of livestock are three-fold. First, there has been an elevation in emissions of methane that is a potent greenhouse gas (GHG) with a global warming potential approximately 25 times that of carbon dioxide. Methane derives primarily from manures and the digestion of cellulose by microorganisms in the rumen of cattle, sheep and other ruminants. Second, eutrophication of rivers and lakes can result from phosphate enrichment derived from animal manures. Finally, and as referred to above, a significant proportion of cultivated land (as distinct from pasture) is used to produce nutrition for livestock.

2.4 Impacts from Fossil-fuel Use

The major environmental manipulations which have allowed the increased genetic potential of crops and livestock to be realised are, one way or another, mostly attributable to the substantial use that has been made of fossil fuel derived energy. Fossil fuel use has impacted significantly in much (but by no means all) of the world's agriculture by facilitating a major increase in productivity. This has radically altered the competition for NPP in man's favour through the ease with which forests can be cleared, land cultivated, crops irrigated and fertilised, predators of livestock reduced and crop pests, diseases and weeds managed. Primary agricultural production consumes, directly or indirectly, approximately 4% of total global fossil fuel; although it must be remembered that perhaps as much as 60% of the food consumed globally is produced by farmers in developing countries who are predominantly reliant on human labour and power provided by water, wind and draught animals. Of the fossil fuel used in global agricultural production, about half is used to synthesise nitrogen fertiliser by means of the Haber–Bosch process.[11]

Agricultural productivity would be far lower than it is today and humans' demand for land much greater (taking account not just of food production but also feed for draught animals) without the supplementary energy provided by the use of fossil fuels (*i.e.* the carbon which was fixed through the photosynthesis that occurred a hundred million years or so ago). In this sense, it can be argued that the environmental impact of agriculture is lessened by fossil energy use or, perhaps more importantly, human depravation and conflicts over land and water resources have been relatively contained. Nevertheless, there are two particular environmental consequences which have followed from the supplementation of energy inputs into food production by access to fossil fuels. The first relates to the increase in reactive nitrogen and the second to water.

2.5   Reactive Nitrogen

It has been estimated that approximately 55% of the reactive nitrogen in the global nitrogen cycle now originates from chemical synthesis as distinct from lightening and biological fixation.[11] This means that many man-managed agro-ecosystems have the potential for adverse environmental consequences due to the leakage of significantly elevated levels of reactive nitrogen as soluble nitrates and gaseous nitrous oxide.[12] The latter is a potent GHG with a global warming potential about 298 times that of carbon dioxide and is produced during the natural processes of nitrification and denitrification. Nitrate enrichment of water and terrestrial ecosystems can, respectively, cause eutrophication and impact adversely on floral biodiversity by encouraging proliferation of rapidly growing and highly competitive species. However, it can also be argued that nitrogen enrichment of natural ecosystems such as grasslands and forests might be expected to increase carbon capture, biomass productivity and thereby reduce atmospheric carbon dioxide concentrations with a mitigating effect on global warming (despite the decline in biodiversity).

Discussion of the fluxes of GHGs from soil (both carbon dioxide and nitrous oxide) places a further emphasis on the importance of land area under different usage and management (the primary preoccupation of this paper). For example, there are large stocks of carbon stored in soil as organic matter under permanent pastures. Trees and forest soils represent equally important stores of fixed carbon. Conversion of grassland or forest to cultivated land for crops results in a release of stored carbon (as carbon dioxide) to the atmosphere with potential adverse environmental impact.[13,14] This suggests that crops should be confined to as small an area as is feasible to deliver the levels of production required to meet demand. However, high yields of crops are only achievable if they have access to a sufficiency of reactive nitrogen in the form of ammonium or nitrate ions. Regardless of origin (biologically fixed or applied as manufactured or organic fertiliser) reactive nitrogen both enables plant growth and is the source of nitrous oxide. There is still much to be learnt about nitrous oxide emissions from soil, but there is likely to be a positive (although not necessarily linear) association between the levels of available nitrogen in soil and the propensity of the soil to emit nitrous oxide. This will be particularly evident when soil water status is favourable for denitrification or at times when available nitrate exceeds crop demand.[15] It follows that the larger the area of either improved grassland (for livestock production), or cropland required to meet food requirements, the larger are likely to be the gross emissions of associated nitrous oxide. However, the emissions per unit of production will be determined by productivity; *i.e.* the efficiency with which the land (and available nitrate) is utilised.

2.6   Water – Excess and Shortage

Turning to water, it is estimated that almost 70% of the water extracted from rivers, lakes and reservoirs globally is used for agricultural production. However, climate change coupled with increased demand and competition for supplies from domestic and industrial users are likely to result in potentially serious regional shortages over the coming decades.[16] The only reason why substantial populations of people can live and farm in extremely arid areas of the world is because it is possible to capture and store water as well as to move it fairly readily provided there is energy for construction of reservoirs and canals, for pumps and for irrigation equipment. All this effort to manage water is for the benefit of agricultural production and is often in competition with the needs of both aquatic and terrestrial ecosystems that depend on a sufficiency of water, as well as natural seasonal fluctuations. Interference with hydrology and hydrological cycles as well as over-extraction or inappropriate irrigation practice leading to salinisation of soil can often have adverse consequences for biodiversity in aquatic and terrestrial environments, including estuaries and wetlands valued as wildlife habitat.[17] It is worth noting, however, that unlike fossil fuel, water is not an exhaustible resource when considered globally; the problem is that there is often too much or too little of it in the wrong place at the wrong time and there is some evidence that extreme weather events which provoke these circumstances are increasing in frequency. This is interpreted by some to be a consequence of anthropogenic global warming.[18] It is an absolute necessity for agriculture that water is managed effectively to avoid flood damage and mitigate drought, as well as to drain soils prone to water-logging and irrigate crops growing in arid conditions (including parts of the UK). For this reason, removing or reducing the inevitable environmental consequences, beyond agriculture, that are associated with water surplus and scarcity are perhaps the most difficult of all to effect. Nevertheless, in the UK at least, a great deal of benefit could be derived from appropriate investment in engineering solutions to harvest, store and deal with excess water.

2.7   Contaminants and Pollutants

Apart from the focus above on GHG emissions and impacts on biodiversity relating to fossil fuel use, water availability and land management, there are a few other environmental consequences resulting from agriculture which need to be noted. Agricultural chemicals, such as synthetic pesticides and antibiotics, can be damaging to non-target organisms or enter food chains with adverse effects. Advances in ecotoxicology, regulation and user training have markedly reduced these impacts in recent decades but one striking contemporary example of unanticipated environmental consequence was the mortality and population decline of vultures in India. This is thought to be a consequence of their feeding on livestock carcasses contaminated with the veterinary medicine, diclofenac.[19]

In a different context, it is also now well recognised that phosphate applied as a fertiliser (or from livestock manures) can cause serious eutrophication of water when washed off the soil surface into rivers, carried in eroded soils or by leaching after unnecessary over-provision to soils that are already phosphate-sufficient.[20] Contamination of water (which may subsequently be destined for drinking) by pesticides is also often cited as an environmental problem and certainly can be when, infrequently, there are large and catastrophic discharges.[21] However, in the normal course of events when modern pesticides are used in accordance with manufacturers' recommendations, environmental impacts are either undetectable or minimal and this will not be considered further.[22]

2.8 Avoiding Negative Environmental Consequences of Agricultural Practice – A Summary

In summary, for reasons that have been elucidated above, there are inevitable environmental consequences of agriculture, and always have been. However, there are ways in which negative environmental impacts can be mitigated and these fall into two categories but both relate to efficiency of resource use. Firstly, developing and favouring systems of production that enable more nutritional output (in terms of calories, protein *etc.*) per area of land or volume of extracted water relaxes the pressure on these resources and enables them to be managed (equally intensively perhaps) for purposes other than food production (wild life conservation, preservation of valued landscapes, carbon capture and storage, bioenergy crop production, recreation *etc.*). Secondly, developing and favouring systems of production that result in the lowest possible net GHG emissions per unit of nutritional output (striving towards the irreducible minimum emissions that the natural processes of well-managed nitrogen cycling; enteric digestion of cellulose; and soil respiration will allow). The first is designed to impact locally and is intrinsically easier to measure progress towards than the second that will primarily have global impact; and will impact locally only in so far as it also involves reductions in use of land and external inputs of potential environmental pollutants. It follows that systems of crop and livestock production to be discouraged will be those that are founded on the utilisation of unnecessarily expansive and unproductive land areas; profligate water extraction from wells and rivers; or result in very low units of production per Kg CO_2 equivalents emitted. This is consistent with the concept of "Sustainable Intensification" which was brought to prominence in a recent report by the Royal Society[23] and is considered more fully below.

3 Sustainable Intensification

3.1 Global Land Use

Of the 13 billion hectares of land on earth, approximately two thirds are able to support significant plant life (*ca.* 9 billion hectares). Of this "green" land area, about 18% (1.6 billion hectares) is used to produce crops and 38% (3.4 billion hectares) is used to graze livestock; the remaining 44% (3.9 billion hectares) is forest, savannah and other types of vegetation of which it is estimated that only about a half is genuinely "wild" and unmanaged.[8,24]

Although there is potentially productive land that is not being managed for food production as efficiently as it might be, and there is also degraded land that was formerly productive which could be restored, there is little new land available that can be used for crop production without reducing areas of permanent grassland, forest or other areas valued either as important carbon sinks and stores or as reservoirs of biodiversity. As concluded from the *Foresight* study,[1] this means that the projected increased demand for food will need to be met from the same (or possibly less) land of which grassland for production of ruminants will remain significant. However, how the latter might best be managed efficiently to produce meat and dairy products (in terms of numbers of animals and associated emissions of methane and nitrous oxide) while also maximising capture and storage of carbon in soil is an important subject for future research.

The concept of Sustainable Intensification acknowledges that land availability is a constraint and that "extensification" of low-productivity agriculture cannot be the solution to sustainable global food production. The concept acknowledges that the primary objective of land use for agriculture is the efficient conversion of solar energy into varied and valued forms of chemical energy. It also acknowledges that some land might best be used to produce forage for animals as intermediates in the energy conversion process. The intensification of land use is aided by a focus on optimising the manipulation and management of the interaction between genotype (livestock and crop) and the environment. The implicit characteristics of a sustainable system (persistence, resilience, autarchy and benevolence)[23] are captured by the recognition that attention must be paid to the continuity of agro-ecosystem functions such as nutrient cycling, pollination and biological buffering of pest and pathogen populations. Pro-active management of the agro-ecosystem has been referred to as "ecological intensification" which captures the concept of a system founded on deep knowledge of relevant ecological science and interventions with reliable and predictive outcomes.

A logical conclusion is that maximising efficiency on the smallest necessary land area provides options to use non-agricultural land (*i.e.* land spared from agricultural production even if in close proximity) to achieve other objectives (environmental, cultural and social). Such objectives should not be confounded with the requirement to produce food and other agricultural products on the land best suited for the purpose and as efficiently as possible in terms of resource use and emissions.

3.2 Anthromes and Anthropogenic Ecosystem Processes

In 2008, Ellis and Ramankutty introduced the concept of "anthropogenic biomes" (now referred to as "anthromes"; http://ecotope.org/anthromes/paradigm/) as an advance in thinking founded on the classical ecological classification of land as "biomes".[25] The notion has been that biomes are a classification of natural systems (with humans disturbing them) reflecting a function of climate, terrain and geology as the basis for variation. In contrast, anthromes are a classification of human systems, with natural systems embedded in them, and variation between anthromes is a function of anthropogenic ecosystem processes including: population density, how land and resources are used (influenced by affluence and technology) and unknowns still under investigation. The underlying point is that humans control ecosystem functions and biodiversity, as much as climate, through activities such as: deforestation, habitat fragmentation, grazing, arable cropping, urbanisation *etc*. Under the anthrome classification, the following sorts of attributes are taken account of: population density, land use, plant cover, NPP, biodiversity (native and introduced species), GHG emissions, reactive nitrogen *etc*.

It would be a worthwhile venture to start considering Britain as a collection of interlinked anthromes with different characteristics and utilities (in terms, for example, of ecosystem service delivery). It might then be possible to go on to consider what the optimum management for each anthrome might be, in order to best deliver the ecosystem services for which the anthrome is most suited and which collectively could deliver the diversity of economic and environmental outcomes being sought including, of course, a secure supply of food.

The anthrome concept is well aligned with the notion of Sustainable Intensification and the need to ensure food security for a growing world population. The concept rejects the notion that sustainable management has somehow got to minimise human influence; but instead, it acknowledges that humans have reshaped ecosystems and will necessarily be "permanent managers of the biosphere". Such management may include various forms of active influence over human demand for certain types of food, other forms of biomass, and systems of production.

3.3 Examples of Sustainable Intensification

A recent global analysis of the changes in areas and yields of 174 crops in the two decades between 1985 and 2005 indicates where production increases have resulted from increased productivity on substantially the same land area or have come from an expansion in land use.[26] Wheat, rice and cotton are examples of the former while the oilseed crops: sunflower, rape seed and soya bean come into the latter category.

Wheat production in the UK provides a good illustration of Sustainable Intensification both in terms of what has been achieved and could potentially be achieved in future. In the thirty year period from 1965 to 1995, average yields doubled from approximately 4 to 8 tonnes per hectare but have remained substantially unchanged since.[24,27] This is despite the fact that breeders have steadily increased the genetic potential such that, in trials, the highest yielding cultivars will regularly produce 12 tonnes per hectare and some farmers have achieved up to 15% more,[28] with the world record for yield of 15.64 tonnes per hectare having been obtained in New Zealand with a British-bred cultivar ("Einstein"). The disparity between genetic potential and actual yield achieved is defined as "yield gap" and may result from sub-optimal management including the timing or quantities of fertiliser and fungicide applications (possibly based on a determination of the market-derived economic optimum). However, if the yield gap is compensated for by an increased area of cultivation, this could translate into up as much as 50% more cultivated land with all the potential environmental impacts elucidated above.

The importance of optimum nitrogen fertiliser use and effective disease control in wheat with regard to GHG emission per unit of production has been well demonstrated in an analysis of data from nine wheat crops.[29] The GHG emissions from crops receiving optimum fertiliser and fungicide applications were 417 Kg CO_2 equivalents per tonne compared to 740 Kg CO_2 equivalents per tonne for crops experiencing a 25% yield loss in the absence of effective disease control; this 44% greater level of GHG emission per unit of production includes the extra land use required to substitute for lost yield.

It can of course be correctly argued that all materials which depend on fossil fuels for their production (and application) represent unsustainable inputs. Nitrogen fertilisers and fungicides are no exception. However, with the level of understanding that now exists about the molecular identity of plant resistance genes, and the ease with which they can be cloned and manipulated by advanced biotechnology, it could in principle make dependence on chemical disease control unnecessary. Similarly, there are already proven genetic technologies for invertebrate pest control (such as transgenic crops carrying a range of insect toxin genes derived from *Bacillus thuringiensis*) and others (such as crops engineered to produce semiochemical insect repellents) are under development. At the same time, the relatively modest amount of global energy utilised to synthesise nitrogen fertilisers could readily be substituted by renewable sources of electricity and either hydrogen or methane,[30] although the relative cost of energy does not currently make this option economically attractive.

Another good illustration of the way in which sustainable intensification can reduce environmental impact is the elevation of milk yields per cow founded on efficient genetic selection, high health status and optimised nutrition. Reduced methane emissions per litre of milk produced are a consequence of requiring fewer animals with a greater productive longevity to produce the required volumes. Efficient feed conversion also translates into lower land requirements for forage and feed supplements.[31] However, as an illustration that there are always likely to be trade-offs, highly selected dairy herds result in male calves less well configured to supply beef and more suckler cows are required to compensate for this in order to meet demand.

3.4 Management of Biodiversity: Land Sharing or Land Sparing?

As has recently been pointed out, the word "intensive" has meaning only when qualified and is dependent on context.[32] In the context of land management, whether for food production or conservation of biodiversity, the context focuses on intensive application of knowledge, management practices, resource inputs and outcomes that are frequently measured in terms of land areas under management. Hence, biodiversity might be measured in terms of the number of species per unit area or the abundance of a particularly valued species per unit area. In Britain, there has been encouragement for farmers to engage with so-called "environmental stewardship" schemes which deliver a source of income in return for delivery of purported environmental benefits, including improvement of habitats for wildlife on farms by removing land from production. This can be described as a "land sharing" strategy for wildlife conservation which seeks to substitute agricultural production with environmental benefit often by reducing productivity.

Implicit in the Sustainable Intensification of agriculture is the notion that land spared from more extensive and less productive systems will be used to achieve other desirable outcomes. This is described as a "land sparing" strategy since it allows the two functions of increased agricultural productivity and biodiversity conservation, for example, to receive sole attention and, potentially, intensive management. There is a growing body of evidence that a "land sparing" as distinct from a "land sharing" approach delivers better outcomes in terms of biodiversity conservation and, as the examples provided above testify, the same is invariably true for agricultural production.[33,34]

4 Land Use, Resource Management and Deliverables from Land

4.1 Understanding Interactions and Trade-offs

The concept of man-managed ecosystems or anthromes introduced above recognises that there are a number of competing outcomes that are required from land which may be independent of, or closely associated with, the use to which land is put. In matching land use with required outcomes there are a number of resources that are available and through which management can be exercised. This is illustrated in Figure 1 which seeks to point out that it may be possible to analyse and understand the interactions that take place between individual combinations of components (for example for a crop, the energy inputs and the sought-after outcome in terms of yield increase). However, the complex of interactions within the whole system of different land uses, required outcomes and available resources can only be resolved by seeking a high level of quantification; defining as closely as possible the limits of acceptability for particular outcomes to be achieved; and taking a systems-based approach to the analysis of the inevitable trade-offs that will be necessary to arrive at a satisfactory solution.

4.2 Units of Accounting

The units of accounting that are traditionally used to measure productivity or efficiency (*e.g.* tonnes of grain per hectare) fail to take account of either the purpose of production (*e.g.* nutritional value, co-products *etc.*) or the environmental consequence of the means of production (*e.g.* land, energy or water used). Efficiency is a function of the genetic potential of the crop or livestock, the man-managed environment created and some measure of environmental

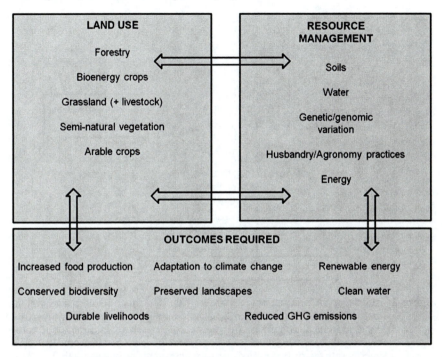

Figure 1 The different outcomes required from land are dependent on the multiple interactions between options for land use and the way available resources are managed. Optimisation of the system requires the quantification and definition of an acceptable set of outcomes and analysis of the system to identify the trade-offs that will be necessary to achieve this.

consequence or cost. This is illustrated in Figure 2. While economists are able to integrate many features of the economics of a system in terms of Total Factor Productivity, there is, as yet, no integrative metric which takes total account of the biophysical productivity of an agricultural system including its environmental impact(s). Using the common language of monetary value and exchange rates to equate quite different entities works well in conventional economic accounting. Attempts have been made to place an economic value on provision of ecosystem services for which there is no market in operation[35] and a value for carbon exists (however imperfect this might be). However, deriving a meaningful economic value for cultural aspects of biodiversity for example is fraught with difficulty and adjusting values by economic devices (such as subsidy or taxation) can discount scientific evidence even if it is easy to effect. For the time being it is probably preferable to remain with a series of different metrics which describe the reality of a man-managed ecosystem while acknowledging the difficulty of dealing with shifting values of different components (water, food, biodiversity *etc.*) at different points in time and under different circumstances.

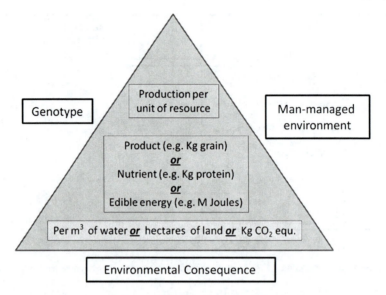

Figure 2   Efficiency of production or productivity is a function of the interaction between genotype (of crop or livestock) employed, the environment created by management and the environmental consequence. This interaction can be described by different metrics dependent on circumstance and no single metric satisfactorily describes performance in terms of sustainability. Hence production of a product (*e.g.* Kg of grain) may be less informative than of a specific nutrient (*e.g.* Kg protein) or energy consumed (*e.g.* MJoules). Efficiency may be defined in terms of resource use (*e.g.* water or land or energy) or environmental consequence (*e.g.* waste or net GHG emissions).

4.3   Some Examples of Trade-offs

Everyone would like to be able to identify "win-win" positions where, for example, an increase in livestock productivity can be made at the same time as having a reducing environmental impact. In the context of the subject of this paper, "win-wins" are probably illusory or at least very rare when circumstances are analysed in depth. The example of pasture management serves to illustrate the point.

Grazed pastures create valued landscapes and land for recreation; a minimum stocking density is required to avoid the encroachment of undesirable vegetation. Such pastures capture and store carbon in the soil and if they are nutrient poor and unimproved, they may also harbour a high floral diversity. Invertebrate and vertebrate biodiversity may also be encouraged by the form of management deployed to maintain the landscape. However, the nutrient value of this unimproved pasture for livestock will be poor and the emissions of methane per unit of meat or milk produced will be comparatively high. More animals will be required to deliver the nutrients and energy to the human diet than would be the case on a pasture managed for high productivity of forage. Improvement of the productivity of the pasture by addition of nutrients would lower floral biodiversity but would likely increase below-ground carbon storage and might even increase quantities of invertebrates, if not diversity. The nutrient enrichment would be likely to increase biomass of vegetation in surrounding areas such as hedgerows and thereby capture more carbon here also. The same area of intensively managed pasture would produce more milk or meat from fewer animals and the higher quality diet would result in lower methane emissions per unit of production. It may be, however, the water courses in the vicinity of the intensively managed pasture would be more at risk from nutrient enrichment from run-off and animal manures.

The scenarios painted above do not have numbers attached to them and quantification is essential for sound decision making, However, this rather simplistic outline makes the point that there are trade-offs everywhere between different trophic levels of biodiversity, carbon capture and storage, net GHG emissions, land areas deployed and productivity of human nutrients. Determination of the boundaries for acceptable outcomes together with good data and sound analysis is the only way of quantifying the trade-offs and deciding on the course of management to be deployed; and this will of course be markedly influenced by the economic inducement or deterrence which the market or policy framework dictates is appropriate. Systems analysis must become a fundamentally important tool if there is to be a well-founded balancing of the environmental consequences of agriculture and the need for food security although it is recognised that there are economic and social factors that will inevitably influence the ready implementation of policies founded purely on a biophysical analysis of the sort indicated above.

5   A Systems-based Approach to GHG Balance

5.1   *Agriculture as Part of the Problem and Part of the Solution*

Agricultural production is a significant source of GHG[36] which were estimated by the UK Department for Energy and Climate Change (DECC)[37] in 2011 to be 49.5 Mt CO_2 eq. p.a of the 566.3 Mt CO_2 eq. p.a. of UK emissions (*i.e.* 8.7%). These emissions are primarily of methane (36%) and nitrous oxide (55%) with carbon dioxide being a minor component (8%). The 2008 UK Climate Change Act commits government to achieving a reduction of 80% in GHG emissions by 2050 based on the estimated 1990 baseline of 748 Mt CO_2 eq. Agriculture is required to play its part in achieving this ambitious target and the route by which this might be achieved has been proposed together with the associated costs.[38] In response, a group of more that twelve influential bodies representing the interests of the UK agricultural industry have come together to develop an industry-led "GHG Action Plan" designed to avoid the requirement for regulation.[39] At the commencement of the process, an agreement was reached with government that on-farm actions would be identified and promoted that would be likely to result in the reduction of 3 Mt CO_2 eq. p.a by 2020 that was set as a target. There were fifteen such actions agreed in the

Action Plan which are also likely to reduce producer costs since they primarily reflect an increase in resource use efficiency.

While the actions to be promoted in the Action Plan will undoubtedly be beneficial, it is questionable whether the setting of emissions reduction targets is at all helpful for the agriculture (or other land-based) sector. In all other area of the economy, GHG emissions reduction is substantially about reduction of CO_2 emissions by reducing fossil fuel consumption and moves to alternative sources of energy; this is easily measurable. Agriculture will benefit from any moves towards reducing reliance on fossil fuels but this is a minor component of the GHG emissions associated with crop and livestock production. Reductions in methane and nitrous oxide will be required to impact significantly on net GHG emissions from agriculture and yet there is enormous uncertainty about the actual levels of baseline emissions, as well as the feasibility of managing the natural biological processes that give rise to these particular GHGs. These emissions are in some measure proportional to the scale of production of crop and livestock products which makes the task of seeking reductions all the more difficult when higher levels of production are being sought.

5.2 Fossil Fuel Substitution, Carbon Capture and Storage, Food Imports and the Cost of Valued Landscapes

Rather than setting and seeking to meet what are arbitrary time-bound GHG emissions reductions, it would perhaps be more appropriate to set about calculating, as accurately as possible, what the irreducible minimum emissions of methane and nitrous oxide might be from UK production systems designed to be highly efficient and capable of producing the volumes and diversity of indigenous foodstuffs required to meet the likely needs of a projected human population of 70 million. Policy implementation would then focus on the inducement to deliver the levels of efficiency demanded in terms of land and other resource use; all this would be consistent with the Sustainable Intensification concept.

Several other things follow from this approach. The only well-proven means of carbon capture and storage is through photosynthesis and subsequent sequestration as organic matter in soil (including peat) and woody biomass. In addition to deriving the irreducible minimum figure referred to above, it should also be possible to derive a figure for the annual carbon capture and storage potential of appropriately managed land (grassland, forests, parks, gardens *etc.*) which has been "spared" from inefficient food production. The range of land-based approaches to the generation of renewable energy as a substitute for fossil fuel use could also be taken account of and encouraged (or otherwise) as appropriate. All this would allow a calculation to be made of how far out of balance the whole UK man-managed land system was in terms of emissions, carbon capture and fossil-fuel substitution.

It is of course highly unlikely that the system would be in balance; the dependency of the UK on "imported" ecosystem services has recently been well elucidated.[40] However, with increasingly refined data and analyses this systems-based approach would allow policies to be implemented that related not just to agricultural production and land management but also food (and other biomass) importation and land use for bioenergy crops. There would be logic in the implementation of policies to discourage the importation of food from production systems that were evidently less efficient in terms of GHG emissions per unit of product than UK derived products. Similarly, there would be sense in discouraging production in the UK of food products that could be imported from countries where the systems of production were demonstrably more or equally as efficiently as was the case at home. This policy might allow more UK land resource to be allocated to delivering carbon balance. If the latter was, despite all efforts, unachievable by sourcing food from outside the UK then a policy of "paying" for an appropriate level of carbon capture by encouraging appropriate land-use practices outside the UK would be indicated.

A rider to these arguments is that a carbon "cost" of maintaining valued (and valuable) UK landscapes and ecosystems that might be producing significant net GHG emissions would also have to feature in the accounting process. In this way, it might be possible to begin to value (and pay for), in the biophysical currency of CO_2 equivalents, the biodiversity and societal benefits that are derived from the conservation of particular land usage (such as extensive grazing on unimproved upland grasslands for example) when food production is evidently not "economic" when defined in these terms.

6 Conclusions

A number of themes have emerged from this wide ranging, but far from comprehensive, treatment of subjects that have major relevance to the current intense interest in how it may be possible to achieve a better balance between the environmental consequences of agriculture and the need for food security. These are as follows:

i. There is need to understand better how agricultural systems have developed over time and continue to develop; why they work well when they do so and, equally, why they fail to deliver on some of the essentials that define sustainability.
ii. The concept of Sustainable Intensification is powerful, and the following is a hypothesis that should be rigorously tested: the most environmentally benign way to produce food is to deliver what the market requires, as efficiently as possible in resource-use terms and from the smallest necessary land area.
iii. More and better data are rather urgently required about many different aspects of the functioning of agricultural systems; several different types of model to describe and probe such systems have been eluded to and, in a UK context, with economic and environmental benefits in mind, there is lots of scope for innovation and adoption of new approaches to the development and implementation of robust policies.

iv. Although unpalatable to some, it is necessary to accept that there are few
    if any "win-wins" to be had and, if this is accepted, the priority is to be
    able to identify and quantify with as much certainly as possible the
    necessary trade-offs such that acceptable, but compromised, outcomes
    can be arrived at which may differ dependent on time, place and priority
    for the land manager or owner.

References

1. *Foresight. The Future of Food and Farming*, Final Project Report, The
   Government Office for Science, London, 2011.
2. *UK National Ecosystem Assessment, Synthesis of the Key Findings*, UNEP-
   WCMC, Cambridge, 2011.
3. H. Stace and J. G. Larwood, *Natural Foundations: Geodiversity for People,
   Places and Nature*, English Nature, Peterborough, UK, 2006.
4. M. W. Bruford, D. G. Bradley and G. Luikart, *Nature Rev.*, 2003, **4**, 900.
5. J. F. Hancock, *Plant Evolution and the Origin of Crop Species*, CABI
   Publishing, Wallingford, UK, 2004.
6. W. C. Lowdermilk, Conquest of land through seven thousand years, *U. S.,
   Dep. Agric., Bull.*, 1953, No. 99, revised 1975, reprinted, 1994.
7. C. Pontin, *A New Green History of the World: The Environment and the
   Collapse of Great Civilisations*, Penguin, New York, 2007.
8. E. C. Ellis, K. K. Goldewijk, S. Siebert, D. Lightman and N. Ramankutty,
   Global Ecol. Biogeog., 2010, **19**, 589.
9. H. Haberi, K.-H. Erb and F. Krausmann, Global human appropriation
   of net primary production (HANPP), in *Encyclopedia of Earth*, ed. C. J.
   Cleveland, National Council for Science and the Environment, Washington DC, 2010.
10. J. A. Burney, S. J. Davis and D. B. Lobell, *Proc. Natl. Acad. Sci. U. S. A.*,
    107, 12052.
11. V. Smil, *Enriching the Earth – Fritz Haber, Carl Bosch, and the Transfor-
    mation of World Food Production*, MIT Press, Massachusetts, 2001.
12. C. Brink, P. Smith, D. Martino, Z. Cai, D. Gwary, H. Janzen, P. Kumar,
    B. McCarl, S. Ogle, F. O'Mara, C. Rice, B. Scholes, O. Sirotenko, M.
    Howden, T. McAllister, G. Pan, V. Romaneukov, U. Schneider, S.
    Towprayoon, M. Wattenbach and J. Smith, Costs and benefits of nitrogen
    in the environment, in *The European Nitrogen Assessment – Sources, Effects
    and Policy Perspectives*, Cambridge University Press, Cambridge, 2011.
13. P. Smith, *et al.*, *Philos. Trans. R. Soc. London, Ser. B*, 2008, **363**, 789.
14. N. Fitton, C. P. Ejerenwa, A. Bhogal, P. Edgington, H. Black, A. Lilly,
    D. Barraclough, F. Worrall, J. Hillier and P. Smith, *Soil Use Manage.*
    2011, **27**, 491.
15. F. Conen and A. Neftel, Nitrous oxide emissions from land-use and land-
    management change, in *Nitrous Oxide Climate Change*, ed. K. Smith,
    Earthscan, London, 2010.

16. K. Strzepek and B. Boehlert, *Philos. Trans. R. Soc. London, Ser. B*, 2008, **365**, 2927.
17. P. Humphries and D. Baldwin, *Freshwater Biol.*, 2003, **48**, 1141.
18. S.-K. Min, X. Zhang, F. W. Zwiers and G. C. Hegerl, *Nature*, 2011, **490**, 378.
19. R. E. Green, I. Newton, S. Schultz, A. A. Cunningham, M. Gilbert, D. J. Pain and V. Prakash, *J. Appl. Ecol.*, 2004, **41**, 793.
20. R. W. McDowell, A. N. Sharpley, L. M. Condron, P. M. Haygarth and P. C. Brookes, *Nutr. Cycling Agroecosyst.*, 2001, **59**, 269.
21. P. Dowson, M. D. Scrimshaw, J. M. Nasir, J. N. Bubb and J. N. Lester, *Water Environ. J.*, 1996, **10**, 235.
22. N. M. van Straalen and C. A. M. van Gestel, *Water, Air Soil Pollut.*, 1999, **115**, 71.
23. Royal Society, *Reaping the Benefits – Science and the Sustainable Intensification of Global Agriculture*, RS Policy Document 11/09, London, 2009.
24. FAOStat, 2011; http://faostat.fao.org
25. E. C. Ellis and N. Ramankutty, *Front. Ecol. Environ.*, 2008, **6**, 439.
26. J. A. Foley, N. Ramankutty, K. A. Brauman, E. S. Cassidy, J. S. Gerber, M. Johnston, M. D. Mueller, C. O'Connell, D. K. Ray, P. C. West, C. Balzer, E. M. Bennett, S. R. Carpenter, J. Hill, C. Monfreda, S. Polasky, J. Rockström, J. Sheehan, S. Sieber, D. Tilman and D. P. M. Zaks, *Nature*, 2011, **478**, 337.
27. S. Boulton and I. R. Crute, Crop nutrition and sustainable intensification, *Proceedings International Fertiliser Society*, Leek, UK, 2011, **695**, 1.
28. HGCA, *Winter Wheat Recommended List 2012–2013*, Stoneleigh, UK, 2011.
29. P. M. Berry, D. R. Kindred, J. E. Olesen, L. N. Jorgensen and N. D. Paveley, *Plant Pathol.*, 2010, **59**, 753.
30. S. Ahlgren, A. Baky, S. Bernesson, Å. Nordberg, O. Norén and P.-A. Hansson, *Bioresource Technol.*, 2008, **99**, 8034.
31. J. L. Capper, R. A. Cady and D. E. Bauman, *J. Animal Sci.*, 2009, **87**, 2160.
32. C. Spedding, *World Agriculture – Problems and Potential*, 2011, **2**(1), 36.
33. R. E. Green, S. J. Cornell, J. P. W. Scharlemann and A. Balmford, *Science*, 2005, **307**, 550.
34. B. Phalan, M. Onial, A. Balmford and R. E. Green, *Science*, 2011, **333**, 1289.
35. I. J. Bateman, *et al.*, Economic values from ecosystems, in *UK National Ecosystem Assessment Technical Report*, 2011, Ch. 22, p. 1068.
36. S. A. Montzka, E. J. Dlugokencky and J. H. Butler, *Nature*, 2011, **476**, 43.
37. HM Government, *The Carbon Plan – Delivering Our Low Carbon Future*, 2011.
38. A. Wreford, D. Moran and N. Adger, *Climate Change and Agriculture: Impacts, Adaptation, and Mitigation*, OECD, Paris, 2010.
39. National Farmers' Union, *Meeting the Challenge: Agriculture Industry GHG Action Plan – Delivery of Phase I: 2010–2012*, National Farmers' Union, Stoneleigh, 2011; http://www.nfuonline.com/ghgap/
40. T. Weighell, *et al.*, UK dependence on non-UK ecosystem Services, in *UK National Ecosystem Assessment Technical Report*, 2011, Ch. 21, p. 1046.

Positive and Negative Impacts of Agricultural Production of Liquid Biofuels

LUCAS REIJNDERS

ABSTRACT

Agricultural production of liquid biofuels can have positive effects. It can decrease dependence on fossil fuels and increase farmers' incomes. Agricultural production of mixed perennial biofuel crops may increase pollinator and avian richness. Most types of agricultural crop-based liquid biofuel production, however, have a negative effect on natural ecosystems and biodiversity. Comparisons of fossil fuels and liquid biofuels regarding their life cycle emissions of pollutants which are (eco)toxic or contribute to oxidizing smog, acidification or nutrification give mixed results. Fossil fuels often do better than biofuels as to emissions of acidifying and nutrifying substances, but often worse regarding ecotoxicity. A rapid increase in biofuel production can increase malnutrition due to its upward effect on food prices. Liquid biofuel lifecycles are linked to greater water consumption than their fossil fuel counterparts. Life cycle greenhouse gas emissions of liquid agricultural biofuels are currently often larger than those of their fossil fuel counterparts, when effects of land use on carbon stocks are included. An exception in this respect is current Brazilian sugarcane ethanol. When soil quality is to be maintained, there seems to be little scope to convert lignocellulosic harvest residues into liquid biofuels. Agricultural biofuels are much poorer converters of solar energy into usable energy than photovoltaic cells.

Issues in Environmental Science and Technology, 34
Environmental Impacts of Modern Agriculture
Edited by R.E. Hester and R.M. Harrison
© The Royal Society of Chemistry 2012
Published by the Royal Society of Chemistry, www.rsc.org

1 Introduction

One of the functions of current agriculture is the production of biofuels. Agricultural biofuel production mostly regards liquid biofuels, such as bio-ethanol and biodiesel, but biogas (methane) and solid fuels are also produced.[1] In this chapter, the focus will be on liquid biofuels. Current liquid biofuels are mostly based on edible parts of food crops, such as sugarcane, sugar beet, corn (maize), wheat, oil palm, soybeans, rapeseed and sunflower.[1,2] There is some liquid biofuel production based on crops which generate inedible oils, such as *Jatropha* and *Ricinus*, however, and there is experimental liquid biofuel production based on lignocellulosic agricultural produce.[1,2] Worldwide, about 1% of all cropland serves the production of biofuels,[3] but the area under biofuel crops is to increase rapidly when current biofuel policies are implemented.[1,4]

As edible parts of food crops are the main basis for liquid biofuel production, the positive and negative impacts thereof match to a considerable extent the corresponding impacts of food production. But there are also some added dimensions to the impacts of current biofuels. Firstly, the crops used for biofuel production are, for financial reasons, commonly preferentially produced on good quality soils and use the same resource inputs (such as fertilizers and water) as food crops to generate produce.[5,6] This increases competition for scarce resources. Also, food consumption is relatively inelastic and especially a rapid expansion of liquid biofuel production based on edible parts of food crops may not be matched by an increased production on existing agricultural soils.[7,8] This context for biofuel production in practice leads to an upward pressure of rapidly expanding liquid biofuel production on prices of resources used in food production and on food prices and to the replacement of nature by agriculture.[1,6–8] Secondly, energetic aspects matter more in biofuel production than in the case of food production. In biofuel production, a wide margin between energetic inputs and outputs, or a high energetic return on energy investment (EROI)[9] is to be preferred (see also section 3). Consequently, comparisons involving biofuels should be made with other fuels rather than with foods.

Biofuels made from edible parts of food crops have been much criticised and there have been calls for the use of lignocellulosic, crops as a basis for biofuel production.[10,11] However, it may be noted that biofuel yields from those crops are not necessarily larger than from current starch and sugar crops, including their lignocellulosic parts, whereas under market conditions lignocellulosic crops are likely to compete for good soils and inputs which could also be used for food production.[11] The latter might well have an upward effect on food prices.[11] Against this background, the focus of this chapter will largely be on biofuels currently produced from food crops. Sections 3 to 8 will discuss aspects of liquid biofuels generated from edible parts of food crops. In section 7 the production of biofuels from food crop-based harvest residues will be considered.

Biofuels have a variety of impacts which are summarised in Box 1, and which will each be considered in turn. The sections dealing with these impacts will be

Box 1   Impacts of current agriculturally produced biofuels to be considered in
this chapter.

Energy
    – net solar energy conversion efficiency
    – replacement of fossil fuels

Water footprint
Greenhouse gas balance and carbon debt
Emission of pollutants
Impacts on living nature (biodiversity, ecosystem quality and ecosystem
    services)
Food prices, hunger

concluded with an evaluation of their character, be it positive or negative.
Among the impacts are the energetic and environmental impacts. To evaluate
such impacts, as will be further explained in section 2, a 'life-cycle' view,
starting with the seed and ending with combustion-linked emissions, is often
considered preferable.

2   Agricultural Production as Part of Biofuel Life Cycles and Life Cycle Assessment

Products, such as biofuels, have life cycles, of which primary production, such
as agricultural production, is a part. Life cycle assessments (LCAs) consider
biofuels from 'cradle to grave', in the case of liquid biofuel for transport this
means 'from seed to wheel'. Life cycle assessments are often considered the
proper way to compare the energetic and environmental performance of specific biofuels with other fuels or types of energy supply.[1,7,8]
    Life cycle assessment is generally divided into four stages:[12]

    i. *Goal and Scope Definition*;
    ii. *Inventory Analysis*;
    iii. *Impact Assessment*; and
    iv. *Interpretation*.

In the (i) *Goal and Scope Definition* stage, the aim and the subject of life cycle
assessment are chosen. This implies the establishment of 'system boundaries'
and usually the definition of a 'functional unit'.
    A functional unit is a quantitative description of the service performance of
products. It may for instance be: 1 GJ (gigajoule) of biofuel (measured as lower
heating value). This allows for comparing different products providing the
same service.

In the (i) *Goal and Scope Definition* stage, system boundaries are drawn between technological systems and the environment, between significant and insignificant processes and between technological systems.

In drawing up system boundaries, one should consider the matter of significant indirect effects of products. When carbohydrates or lipids from food crops are diverted to biofuel production, this diversion may give rise to additional food and/or feed production elsewhere, because demand for food and feed is highly inelastic.[7,8,13] This, in turn, may have a substantial impact on food prices and the environment. Melamu and von Blottnitz[14] have for instance pointed out that in South Africa the use of bagasse (a residue from sugarcane processing) for the production of lignocellulosic liquid biofuels may reduce its use as a source of energy in industry and may have the indirect effect of favoring the substitution in industry of bagasse by coal. Also a substantial addition of biofuels to the fuel mix may, dependent on actual government policy, impact transport fuel and fossil fuel prices, which in turn may have an effect on transport fuel and fossil fuel use and associated greenhouse gas emissions.[15] The choice of system boundaries may have a substantial effect on the outcomes of life cycle assessments.[1,16]

The (ii) *Inventory Analysis* gathers the necessary data for all processes involved in the product life cycle and the (iii) *Impact Assessment* stage deals with estimating impacts of the life cycle.

When dealing with the environmental impacts of agricultural products in these stages of life cycle assessment, the problem arises that crops may have more than one output.[1,17] For instance, rapeseed processing not only leads to the output oil, which may be used for biodiesel production, but also to rapeseed cake, which may be processed into rapeseed meal to be used as (animal) feed ingredient. Also, biorefineries produce a variety of product outputs. In the case of multi-output processes, extractions of resources and emissions have to be allocated to the different outputs. There are several ways to do so. Major ways to allocate are based on physical units (*e.g.* energy content or weight of outputs) or on monetary value (price). There may also be allocation on the basis of substitution. In the latter case, the environmental burden of a co-product is established on the basis of another, similar product. Occasionally there is disagreement about the question of whether there should be allocation to any biofuel co-product at all. There are also usually non-product outputs ('wastes'). There is often, though not always, no allocation to such 'wastes', but this may change when the non-product output is found to be useful as an input in a production process, and gets a positive monetary value. On the other hand, co-products may be 'overproduced' to such an extent that they are handled as 'wastes'. This has for instance happened with glycerol, a co-product of biodiesel production.[1] Changes of co-product into non-product output *vice versa* may lead to changes in allocation. Different kinds of allocation may lead to different outcomes of life cycle assessments.[1,14,18]

The (iv) *Interpretation* stage of life cycle assessment connects the outcome of the assessment done to the real world. In this stage uncertainties in outcome can be considered and conclusions can be drawn.

The life cycle environmental assessments of agricultural biofuels performed so far, covering the emissions of pollutants, greenhouse gas balances, water consumption and impact on biodiversity, have shown that the stage of agricultural production is the major determinant of agricultural biofuel impacts.[1,8,19–23]

3 Energy

3.1 Solar Energy Conversion Efficiency of Current Agricultural Crop-based Liquid Biofuels

Producing biofuels is one of the ways to convert solar energy into energy useful for mankind. It is not the only way to do so, however, other examples include solar heaters and photovoltaic cells. The latter convert solar energy into electricity. Taking the lifecycle view, net solar energy conversion efficiencies may be calculated in which the energy input into the product life cycle is corrected for.[1] Here allocation to agricultural co-products will be on the basis of monetary value. The net efficiency with which solar energy is converted into energy useful to mankind is different for different energy conversion technologies and biofuels. This is illustrated in Table 1.

In interpreting this table, one should note that as to the second column the quality of the useful energy evaluated varies. The quality of electricity is relatively high, if compared with the quality of biofuels. Table 1 makes clear that there are major differences in solar energy conversion efficiency.

Solar energy conversion efficiencies are an important determinant of land use.[1] Table 1 suggests for a certain amount of useful energy, biofuel production would require much more land than the generation of photovoltaic energy.

Table 1 Net solar energy conversion efficiencies for a variety of biofuels and photovoltaic cells (calculated from references 1 and 24).

Source	Fuel/ electricity	Type of car traction technology	Estimated net conversion efficiency for solar energy into fuel/ electricity (%)	Estimated net conversion efficiency for solar energy into car kilometers (%)
Oil palm (S.E. Asia)	Palm oil biodiesel	Diesel motor	0.15	0.044
Sugarcane (Brazil)	Ethanol	Otto motor	0.16	0.026–0.035
Wheat (Europe)	Ethanol	Otto motor	0.024–0.03	0.0038–0.0066
Rapeseed (Europe)	Rapeseed biodiesel	Diesel motor	0.034	0.010
Multicrystalline photovoltaic cells	Electricity	Electromotor/ batteries	12.5	3.9–10.5

In Europe one may *e.g.* 'harvest' at least 390 times as many kilometers for automotive transport from a hectare of solar cells than from a hectare of rapeseed. This has important consequences for the impact of biofuel production on living nature (see section 7).

3.2 Replacement of Fossil Fuels

An important benefit of biofuels is that the dependence on fossil fuels may be reduced.

The ability to replace biofuels may be expressed in terms of 'net energy yields per hectare'. In calculating these yields, life cycle cumulative fossil fuel inputs are subtracted from biofuel yields.[1]

In Table 2 net energy yields ha^{-1} from biofuels have been calculated by subtracting the lower heating value of the life cycle cumulative fossil fuel input from the lower heating value of the biofuel produced from a hectare of land. This table also includes a net energy yield for photovoltaic modules in which the lower heating value of the current cumulative fossil fuel input ha^{-1} into the photovoltaic module life cycle is subtracted from the electricity yield in $GJ\ ha^{-1}$.

Table 2 shows that biofuels from tropical crops such as the oil palm and sugarcane do better than European and US starch crops, but that multicrystalline photovoltaic modules give much higher net energy yields than the biofuels mentioned in Table 2. When the focus is placed on energy services such as car-kilometers, multicrystalline photovoltaic modules do even better, as electric traction has higher energy efficiency than that of the combustion engine.[1]

The ability to replace fossil fuels is a positive aspect of biofuels. From Table 2 it appears that in this respect there are substantial differences between agricultural biofuels. Sugarcane and palm oil tend to do better than ethanol from starch and oil crops produced in the USA and Europe.

Table 2 Net energy yields reflecting the potential to displace fossil fuels in Giga Joules (GJ) per hectare for selected biofuels and photovoltaic modules (calculated from references 1 and 24). The net energy yield is obtained by subtracting cumulative fossil fuel demand from gross energy yield.

Crop	Location	Product	Net energy yield in GJ (lower heating value) $ha^{-1}\ year^{-1}$
Sugarcane	Brazil	Ethanol	160–175
Oil palm	Malaysia	Palm oil	140–180
Starch crops, soybeans, rapeseed	USA, Europe	Ethanol, oil	35–60
Multicrystalline photovoltaic modules	Brazil	Electricity	$76\times10^2 - 87\times10^2$

3.3   Energetic Return on Investment (EROI)

The energetic return on investment (EROI) is the energy delivered divided by the life cycle input of (man-made) energy.[25] A relatively low EROI has negative implications because the more the energy input to make a specified quantity of energy available, the less will be left for non-energy activities.[25] Agricultural biofuels are currently mainly used as transport biofuels.[1] Current major transport energy sources (gasoline, diesel, electricity) have EROIs > 5.[1,26] This also holds for novel sources of electricity such as wind power systems and photovoltaic cells.[26] None of the current agricultural biofuels has an EROI > 5.[1,14,25,27–29] For instance, the main US biofuel, ethanol from corn, has been estimated to have an EROI of 1.08 (with a standard deviation of 0.2).[29]

4   Water Footprints of Current Agricultural Crop-based Liquid Biofuels

The water footprint of a biofuel is the total volume of fresh water consumed to produce a specified amount of biofuel (*e.g.* 1 GJ, defined as lower heating value of combustion). This includes rainwater, surface water and groundwater evaporated during the biofuel lifecycle, and the volume of water needed to dilute pollutants discharged during the biofuel lifecycle to quality standards.[30–34]

Life cycle fresh water consumption of biofuels tends to be dominated by water use during primary production (cultivation of crops).[3,21,30–34] Life cycle fresh water consumption tends to be much larger for biofuels than for their counterpart fossil fuels. For instance, water footprints of corn ethanol or soybean biodiesel exceed those of conventional gasoline or diesel by at least one order of magnitude.[21,35,36]

Water footprints may vary considerably, dependent on source (*e.g.* crop), crop yield, water management (*e.g.* type of irrigation) and type of allocation. Variability in determinants of water consumption may lead to major differences in water footprints, *e.g.* for the biofuel Jatropha oil the water footprint may range between 250 and 1700 m^3 GJ^{-1}, the latter referring to the lower heating value of Jatropha oil.[30] Gerbens-Leenes *et al.*[31] have estimated average water footprints for several liquid biofuels based on data regarding main producers of the crops from which the biofuels are derived (see Table 3).

If the comparison in Table 3 had been made not on the basis of lower heating values, but on the basis of similar 'seed-to-wheel' performance, biodiesel would have done relatively better as the diesel motor is more efficient than the Otto motor in which ethanol is usually combusted.[1]

A study on the biofuel crop corn in the Midwestern United States suggests that evaporation, an important component of the water footprint, will increase on warming linked with climate change.[37] Replacing corn by the lignocellulosic biofuel crops *Miscanthus* or switchgrass would lead to much higher evaporation.[37]

Aggregated at the level of substantial liquid biofuel production in a region, the water footprint can be large.[3,32–34] For instance Yang *et al.* have estimated

Table 3 Estimated average water footprints in m³ GJ⁻¹ for several liquid biofuels.[31]

Biofuel	Estimated average water footprint in m³ GJ⁻¹ (lower heating value)
Biodiesel from soybeans	394
Biodiesel from rapeseed	409
Ethanol from sugar beet	59
Ethanol from sugarcane	108
Ethanol from maize (corn)	110
Ethanol from wheat	211
Ethanol from sorghum	419

that a yearly production of 12 million metric tons of ethanol from corn, as was originally envisaged by the – now rescinded – Chinese government mandate for the year 2020, would have required the equivalent of the yearly discharge of the Yellow River.[32]

Water increasingly emerges as a constraint on agricultural production and yearly water consumption exceeding added yearly supplies is increasingly common.[3,34] The water footprints of agricultural biofuels may therefore be a negative aspect of agricultural production of biofuels.[31–34,38] It has, for instance, been found that growing *Jatropha* for biofuel production in Tamil Nadu can potentially exacerbate conflicts and competition over water access in villages.[38] It should be noted, however, that water footprints vary strongly depending on the crop, as shown in Table 3, whereas there are also large regional differences in water availability.[3,31,33] This also matters for the impacts of biofuel production on natural ecosystems (see section 7).

5 Life Cycle Greenhouse Gas Emissions and the Carbon Debt of Current Agricultural Crop-based Liquid Biofuels

Life cycle greenhouse gas emissions are usually evaluated on the basis of a time horizon of 100 years and direct effects only.[1] In the calculation of life cycle greenhouse gas emissions linked to liquid biofuels derived from crops, several sources of emissions have to be considered to allow for relatively comprehensive estimates. These include:[1,5]

i. Cumulative fossil fuel demand;
ii. The emission of the greenhouse gas nitrous oxide (N_2O) linked to the use of fertilizer in cropping;
iii. The emission of carbonaceous greenhouse gases linked to changes in soil carbon stocks due to cultivation practices (*e.g.* tillage) and handling of crop residues; and
iv. Emissions linked to changes in carbon stock following from land use change. Such land use change can be direct (*e.g.* cutting forest to allow for the establishment of oil palm plantations) or indirect (following from the inelastic demand for food in feed). As pointed out in section 2, there

may also be other indirect effects of biofuel use, affecting life cycle greenhouse gas emissions. Reversion of land use at the end of biofuel programs, as suggested by Delucchi,[33] is not considered here, as biofuel programs tend to be promoted as being sustainable.

When the coverage of greenhouse gases is comprehensive, as indicated above, many current liquid biofuels do not do well, if compared with fossil fuels, whatever the allocation used.[1,7,13,39,40] There is a large probability that current biofuels have a larger life cycle greenhouse gas emission than current fossil fuels when a 20 year period (as recommended by the IPCC) is used for discounting changes in C stocks due to land use change, though it may be noted that allocation based on prices tends to give rise to a somewhat higher lifecycle greenhouse gas emission than allocation based on physical units.[1,7,13,39,40]

There is a large probability that the 'seed-to-wheel' or 'seed-to-tank' emission of greenhouse gases linked to current biofuels is greater than the corresponding 'well-to-wheel' or 'well-to-tank' emission associated with current conventional transport fuels. An exception would seem to be ethanol from Brazilian sugarcane, if the intensity of cultivation is unchanged and direct and indirect effects of sugarcane cropping on land use do lead to clearing of the Cerrado, the world's largest savannah.[40] In this case, the life cycle greenhouse gas emissions of biofuel are somewhat lower than those of conventional gasoline, when the impact of land use change is discounted over a 20 year period. This would not be the case when sugarcane cropping leads to clearance of rainforest, which has been suggested as a likely development in for the near future.[41] In the latter case, ethanol from sugarcane would do worse than gasoline.[1] On the other hand, when additional ethanol production from sugarcane is linked with increased yields of land currently under sugarcane, one would expect ethanol from sugarcane to do considerably better in life cycle greenhouse gas emissions.[1] Since this is apparently the main way in which the production of Brazilian sugarcane ethanol has expanded,[42] life cycle greenhouse gas emissions of this type of ethanol will be considerably lower than those of gasoline.[1] Another current biofuel which might do well regarding life cycle greenhouse gas emissions would be palm oil from oil palm plantations established on abandoned soils which sequestered little C. However, under market conditions this option appears to be unattractive.[43]

Liquid biofuels such as rapeseed-based biodiesel and wheat- or sugar beet-based ethanol, the main biofuels from European soils, currently do worse as to net greenhouse gas emissions than conventional gasoline and diesel, whatever the allocation used.[1] The same holds for corn-based ethanol, the most important liquid biofuel in the USA.[1,7] Biodiesel based on palm oil from plantations for which rain forest has been cleared does worse as to net greenhouse gas emissions than conventional diesel.[44] In general, it is doubtful whether technical changes will allow for a net benefit of current biofuels regarding greenhouse gas emissions, if compared with diesel and gasoline, when, due to direct or indirect effects on land use, forest has to be cleared for this purpose and such clearance is discounted over a 20 year period.[1,45] The same holds true when

Table 4   Carbon debt of land use change to produce biofuels.

Biofuel	Type of land use change	Carbon debt in years	Reference
Sugarcane ethanol	Clearing wooded Cerrado	17	40
Corn ethanol	Clearing US abandoned cropland	45	40
Jatropha oil	Clearing African Miombo Woodland	> 30	46
Palm oil	Clearing rain forest on mineral soil	75–93	44
Palm oil	Clearing rain forest on peat	> 600	44
Palm oil	Converting degraded grassland into oil palm plantation	< 10	44

feedstocks for biofuels are cultivated on peat.[44] On the other hand, intensification of cultivation might strongly reduce land use claims leading to clearance of forests, when biofuel production expands.

An alternative option for assessing the impact of changes in the (agro)ecosystem stock of carbon due to land use change linked to biofuel production mentioned in section 2 is the calculation of the carbon debt.[40] The carbon debt is expressed in years and can be calculated by dividing the change in carbon stock by the amount of fossil carbon replaced by the yearly production of biofuel.[40] Table 4 gives values for the carbon debt for several types of land use change to produce several biofuels.

6   Life Cycle Emissions of Pollutants Linked to Current Agricultural Crop-based Liquid Biofuels

There are several life cycle assessments of current biofuels with allocation based on monetary value or substitution, which deal with pollutant emissions.[1,18,22] Current liquid agricultural crop-based transport biofuels often do worse than their corresponding conventional fossil fuels in the life cycle emission of compounds that contribute to acidification and eutrophication.[19,23] Current liquid biofuels vary much in their life cycle emissions of ecotoxic compounds and of compounds that may contribute to oxidizing smog.[1,23] Regarding life cycle emissions which may contribute to oxidizing smog, ethanol from sugarcane and biodiesel from palm oil tend to do worse than conventional fossil transport fuels, whereas biodiesel from canola or soybean and ethanol from wheat or corn tend to do better.[1,19,23] Most current liquid biofuels give rise to life cycle emissions of lower ecotoxicity than conventional fossil fuels, but the opposite holds for biodiesel made from Malaysian palm oil or Brazilian soybean oil.[1,23]

Especially when biofuel crops are concentrated in specific areas the pollution impacts thereof may be quite substantial. A case in point is sugarcane cropping in Brazil, which includes periodical burning of sugarcane residues. This in turn gives rise to high emissions of air pollutants such as nitrogen oxides, particulate matter and polycyclic aromatic hydrocarbons,[47–51] which in turn lead to an increased occurrence of respiratory diseases, especially among children and the elderly[51] and to an increased lung cancer risk.[50] Another example is the leaching of nitrate from corn cropping for ethanol production in the USA. The 2022 mandate for liquid biofuel production has been estimated to, *ceteris paribus*, lead to an increase in the flux of organic N by the Mississippi and Atchafalaya rivers to the Gulf of Mexico by 10–34%, increasing hypoxia risk in the Gulf.[52] Also groundwater pollution by pesticides is expected to increase due to this mandate.[53,54]

7 Impact of Agricultural Crop-based Liquid Biofuels on Natural Ecosystems and Biodiversity

The impact of agricultural biofuel production on natural ecosystems and biodiversity seems to be mainly determined by the agricultural stage of the biofuel life cycle.[1,19,44] Most studies have focused on the effect of increased demand for land. The relatively poor net energy yields ha^{-1} presented in Table 2 imply that a substantial production of biofuels leads to a large demand for land. To the extent that the demand for land leads to the expansion of agriculture, the amount of land available for natural ecosystems is reduced. Agricultural practices, such as harvesting practices, use of nutrients (fertilizers), use of agricultural chemicals (such as pesticides) and water management, may also impact biodiversity.

Land use change linked to the replacement of rainforest by oil palm plantations has been studied relatively well in terms of its consequences for biodiversity.[44] Biodiversity on oil palm plantations was found to be reduced if compared with tropical rain forest, the reduction being relatively large for species characteristic for primary tropical forests, with smallholdings doing better regarding avian biodiversity than large plantations.[44,55] Also, cutting rainforests may negatively affect ecosystem services linked to the role of forests in cleaning polluted air, sequestering minerals and in the hydrological cycle. The latter is partly due to changes in albedo.[1] Some mitigation of biodiversity loss could be achieved by including remnants of tropical forests in oil palm plantations.[1,44]

Studies available about the expansion of cropping for biofuels in the North America, Brazil and Europe have been limited in scope. Expanding production of annual bioenergy crops (soybean and corn) to meet the 2022 US biofuel mandate may have the direct effect of a significant decline avian richness, whereas expanding mixed perennial biofuel crops (*e.g.* mixed forbs and grasses) to meet this mandate may have the direct effect of significantly increasing pollinator and avian richness.[10,56–58] The scope for the latter depends on the emergence of a large-scale commercial conversion of lignocellulosic biomass into liquid biofuel. On the other hand, there is the possibility that some of the perennials under consideration may emerge as invasive species having a negative effect on biodiversity.[1,47,59] A study regarding the impact of expanding food crop-based biofuel production in North America on arthropods suggests that there may be negative effects on the natural enemies of current pests and found large uncertainties about the expected effects.[60]

A large expansion of sugarcane production, to increase the supply of the biofuel ethanol, may lead to increased fragmentation of the Brazilian Cerrado, which is a biodiversity hotspot, *e.g.* negatively affecting mammalian biodiversity.[61] Substantial expansion of biofuel crop production in Europe has been predicted to significantly decrease the area under semi-natural vegetation,[62] and to lead to more species that suffer from habitat losses than benefit from new habitats.[63]

Water consumption for liquid biofuel production may also have an impact on natural ecosystems: both aquatic organisms and groundwater-dependent terrestrial ecosystems may be affected.[64–66] A worldwide study found a substantial impact on ecosystem quality of water consumption for biofuel production.[65] A study on ethanol production from corn in Minnesota (USA) did show substantial ecological damage due to water consumption, which was geographically unevenly distributed with a cluster of high impact regions near the centre of the state.[66]

All in all, negative effects on natural ecosystems and biodiversity seem to dominate the expansion of current agricultural biofuel production.

8 Effect of Current Agricultural Crop-based Liquid Biofuels on Food Prices and Hunger

Several studies suggest that an especially rapid expansion of agricultural biofuel production will, *ceteris paribus*, lead to increases in the price of food.[1,6,67] An increased linkage of the prices of major biofuel crops to oil prices, when oil prices exceed \$75 per barrel has been suggested.[68] Increases in food prices may have both positive and negative aspects. A positive aspect is that farmers' incomes can increase.[42] In fact this has been an important driver in US and European biofuel policies.[1] On the other hand, increased prices may lead to increased malnutrition of the poor, which is a negative aspect.[1,6]

It has been argued that biofuel policies should prefer smallholder agriculture over agricultural production exploiting the economies of scale, thus increasing incomes of poor people, thereby limiting malnutrition.[42,69] In actual practice, the financial benefits of sugarcane expansion in Brazil do not seem to have trickled down to the poor.[70] The expansion of oil palm cultivation in Indonesia has apparently had a marginalizing effect on smallholders.[71] In the case of *Jatropha* production for liquid biofuel on marginal soils in Tamil Nadu, aimed at improving the income of the rural poor, the opposite seems to occur; it is particularly the poorer and social backward farmers that have been found to have been impoverished by the introduction of *Jatropha* plantations.[38] Due to relatively high certification costs and insufficient institutional capacity, smallholders are also at a disadvantage regarding current palm oil certification schemes if compared with large-scale producers.[72]

9 Liquid Biofuels from Crop Residues

Liquid biofuels may also be produced from lignocellulosic crop residues. Cellulose and hemicellulose present in such residues may be converted enzymatically to ethanol or butanol.[1,73] For this purpose crop residues relatively rich in cellulose and hemicellulose, such as cobs, leaves and husks in the case of corn stover, are preferred[74] Alternatively harvest residues may be thermochemically converted, *e.g.* into synthesis gas which in turn may serve the synthesis of liquid biofuels.[73,76] Catalytic thermochemical conversions of sugars into liquid alkanes have also been demonstrated.[77]As yet, liquid biofuel production from harvest residues is at the pilot/demonstration plant stage, characterised by substantial technological problems and near-term ethanol costs expected to be significantly higher than the costs of ethanol produced from corn starch.[77–83]

Benefits of liquid biofuels from crop residues which have been claimed include: replacement of fossil fuels, mitigating climate change, reduced pollution and absence of upward effects on food prices.[74–76] The expected future performance of factories converting harvest residues into ethanol is a major determinant for fossil fuel replacement by, and for the environmental impacts of, biofuels from harvest residues.[75,84–86] Based on current expectations, the production of liquid biofuels made from corn stover and wheat straw might improve the overall environmental performance of wheat- and corn-based biofuel production.[76,87–89]

However, there is a major problem with biofuels from crop residues. Though crop residues have been classified as wastes,[90] return of such residues to soils is important for soil organic matter (carbon) stocks.[91,92] Adequate soil organic matter stocks are needed for good crop yields; they limit erosion, are conducive to water and nutrient conservation and enhance soil biodiversity.[75,91,92] Crop residue removal leading to lower soil organic matter and nutrient stocks may lead to reduced crop yields, which in turn might have an upward effect on food prices.[91] Other factors impacting soil organic matter stocks are tillage and climate. Soils under tillage need larger inputs of crop residues than no-till soils to prevent a decrease in soil organic carbon and currently often show reductions in soil organic carbon stocks.[82,86] Soils in warm climates need, *ceteris paribus*, higher inputs of crop residues than soils in temperate climates to maintain soil organic matter stocks.[92] In the absence of good data about the relation between soil organic carbon stock and cultivation practices, including handling of harvest residues, it is as yet uncertain whether there is scope for liquid biofuel production from sugar cane and oil palm harvest residues without negatively impacting soil quality. Even in the case of no-till agriculture in temperate climates, only a fraction of the total crop residue produced may be available for biofuel production when soil organic carbon stocks are to be maintained, the actual amount being site specific.[92] The scope for increased removal of harvest residues when residues from liquid biofuel production can be returned to arable soils is as yet highly uncertain, as the compatibility of production residues with maintaining good quality soils is in doubt, because of residue characteristics such as antimicrobial activity.[75,91,93] Also, economic considerations may not be conducive to a return of biofuel processing residues to arable soils. Currently, residues of processing lignocellulosic materials into cellulose for paper production tend to be incinerated, landfilled and/or discharged with wastewater rather than returned to arable soils.[94–98] The organic residues from sugarcane and palm oil processing are usually not returned to arable soils either.[99,100] When the production of liquid biofuel is integrated in biorefineries, it has been argued that for reasons of cost there is little scope for returning processing residues to soils as maximum use should be made from such residues to produce value-added compounds.[73]

When soils are under tillage there seems to be even less scope for the diversion of residues to biofuel production without jeopardizing soil organic matter stocks than in the case of no-till soils.[1,75,91,92] For the North China Plain, which is under tillage and where large amounts of crop residues are currently not returned to soils but used for cooking and heating, there is for instance suggestive evidence that returning larger amounts of crop residues to arable soils will lead to increased crop yields.[101]

10 Conclusions

Agricultural production of liquid biofuels can have positive effects. It can decrease dependence on fossil fuels and increase farmers' incomes. Agricultural production of mixed perennial biofuel crops may increase pollinator and avian richness. Most types of agricultural crop-based liquid biofuel production, however, have a negative effect on natural ecosystems and biodiversity. Comparisons of fossil fuels and liquid biofuels regarding their life cycle emissions of pollutants which are (eco)toxic or contribute to oxidizing smog, acidification or nutrification give mixed results. Fossil fuels often do better than biofuels as to emissions of acidifying and nutrifying substances, but often worse regarding ecotoxicity. A rapid increase in biofuel production can increase malnutrition due to its upward effect on food prices. Liquid biofuel lifecycles are linked to a larger water consumption than their fossil fuel counterparts. Life cycle greenhouse gas emissions of liquid agricultural biofuels are currently often higher than those of their fossil fuel counterparts, when effects of land use on carbon stocks are included. An exception in this respect is current Brazilian sugarcane ethanol. This would change when increased ethanol production of Brazilian sugarcane would directly or indirectly lead to cutting rainforest. When soil quality is to be maintained, there seems to be little scope to convert lignocellulosic harvest residues into liquid biofuels. Agricultural biofuels are much poorer converters of solar energy into usable energy than photovoltaic cells.

References

1. L. Reijnders and M. A. J. Huijbregts, *Biofuels for Road Transport. A Seed to Wheel Perspective*, Springer, London, 2009.

2. P. Lamers, C. Hamelinck, M. Juninger and A. Faaij, *Renewable Sustainable Energy Rev.*, 2011, **15**, 2655.
3. C. de Fraiture, M. Giordano and Y. Liao, *Water Policy*, 2008, **10**(S1), 67.
4. S. Soimakallio and K. Koponen, *Biomass Bioenergy*, 2011; doi: 10.1016/jbiombioe.2011.04.041.
5. S. M. Swinton, B. A. Babcock, L. K. James and V. Bandaru, *Energy Policy*, 2011; doi:1016/j.enpol.2011.05.039.
6. S. A. Mueller, J. E. Anderson and T. J. Wallington, *Biomass Bioenergy*, 2011, **35**, 1623.
7. T. Searchinger, R. Heimlich, R. A. Houghton, F. Dong, A. Elobeid, J. Fabiosa, S. Tokgoz, D. Hayes and T. Yu, *Science*, 2008, **319**, 1238.
8. P. Björesson and L. M. Tufvesson, *J. Cleaner Prod.*, 2011, **19**, 108.
9. B. J. Solomon, *Ann. N. Y. Acad. Sci.*, 2010, **1185**, 119.
10. J. Hill, *Agron. Sustainable Dev.*, 2007, **27**, 1.
11. L. Reijnders, *Biomass Bioenergy*, 2010, **34**, 152.
12. J. N. Guinee, *Handbook of Life Cycle Assessment*, Kluwer, Dordrecht, The Netherlands, 2002.
13. R. J. Plevin, M. O'Hare, A. D. Jones, M. S. Torn and H. K. Gibbs, *Environ. Sci. Technol.*, 2011, **44**, 8015.
14. R. Melamu and H. von Blotnitz, *J. Cleaner Prod.*, 2011, **19**, 138.
15. W. Thompson, J. Whistance and S. Meyer, *Energy Policy*, 2011, **39**, 5509.
16. G. Finnveden, M. Z. Hauschild, T. Ekvall, J. Guinee, R. Heijungs, S. Hellweg, A. Koehler, D. Pennington and S. Suh, *J. Environ. Manage.*, 2009, **91**, 1.
17. B. Brehmer, R. M. Boom and J. Sanders, *Chem. Eng. Res. Des.*, 2009, **87**, 1103.
18. T. Fruergaard, R. Astrup and T. Ekvall, *Waste Manage. Res.*, 2009, **27**, 724.
19. A. Gasapatos, P. Stromberg and K. Takeuchi, *Agric. Ecosyst. Environ.*, 2011; doi:10.1016/j.agee.2011.04.20.
20. S. Kim and B. E. Dale, *Bioresource Technol.*, 2008, **99**, 5250.
21. R. Dominguez-Faus, S. E. Powers, J. G. Burken and P. J. Alvarez, *Environ. Sci. Technol.*, 2008, **43**, 3005.
22. S. C. Davis, K. J. Anderson-Texeira and E. H. de Lucia, *Trends Plant Sci.*, 2009, **14**, 140.
23. R. Zah, H. Boöni, M. Gauch, R. Hischler, M. Lehman and P. Wagner, *Life Cycle Assessment of Energy Products: Environmental Impact Assessment of Biofuels*, EMPA, St Gallen, Switzerland, 2007.
24. N. Mohr, A. Meijer, M. A. J. Huijbregts and L. Reijnders, *Int. J. Life Cycle Assessment*, 2009, **14**, 225.
25. K. Mulder and N. J. Hagens, *AMBIO*, 2008, **37**, 74.
26. I. Kubiszewski, C. J. Cleveland and P. K. Endres, *Renewable Energy*, 2010, **35**, 218.
27. M. Giampetro and S. Ulgiati, *Crit. Rev. Plant Sci.*, 2005, **24**, 365.
28. F. Cherubini, N. D. Bird, A. Gowie, G. Jungmeier, B. Schlamadinger and S. Woess-Gallach, *Resour., Conserv. Recycl.*, 2009, **53**, 434.

29. D. J. Murphy, C. A. S. Hall and B. Powers, *Environ. Dev. Sustainability*, 2011, **13**, 179.
30. W. Gerbens-Leenes, A. J. Hoekstra and T. H. van der Meer, *Proc. Natl. Acad. Sci. U. S. A.*, 2009, **106**, E113.
31. W. Gerbens-Leenes, A. J. Hoekstra and T. H. van der Meer, *Proc. Natl. Acad. Sci. U. S. A.*, 2009, **106**, 10219.
32. H. Yang, Y. Zou and J. Liu, *Energy Policy*, 2009, **37**, 1876.
33. M. A. Delucchi, *Ann. N. Y. Acad. Sci.*, 2010, **1185**, 28.
34. J. M. Evans and M. J. Chen, *Global Change Biol.*, 2009, **15**, 2261.
35. K. Mulder, N. Hagens and B. Fisher, *AMBIO*, 2010, **39**, 30.
36. M. F. Emmenegger, S. Pfister, A. Koehler, L. de Giovanetti, A. P. Arena and R. Zah, *Int. J. Life Cycle Assessment*, 2011; doi:10.1007/s11367-011-0327-1.
37. P. Ariza-Montobbio and S. Lele, *Ecol. Economics*, 2010, **70**, 189.
38. P. V. V. Le and D. T. Drewry, *Proc. Natl. Acad. Sci. U. S. A.*, 2011, **108**, 15085.
39. K. A. Mullins, W. M. Griffin and H. S. Matthews, *Environ. Sci. Technol.*, 2011, **45**, 131.
40. J. Fargione, J. Hill, D. Tilman, S. Polasky and P. Hawthorne, *Science*, 2008, **319**, 1235.
41. R. Walker, *Ann. Assoc. Am. Geogr.*, 2011, **101**, 929.
42. M. Harvey and S. Pilgrim, *Food Policy*, 2011, **35**, 540.
43. B. Wicke, V. Fornburg, M. Juninger and A. Faaij, *Biomass Bioenergy*, 2008, **32**, 1322.
44. F. Danielsen, H. Beukema, N. D. Burgess, F. Parish, C. Brühl, P. F. Donald, D. Murdyarsno, B. Phalan, L. Reijnders, M. Struebig and E. M. Fitzherbert, *Conserv. Biol.*, 2009, **23**, 348.
45. H. K. Gibbs, M. Johnston, J. A. Foley, T. Holloway, C. Monfreda, N. Ramankutty and D. Zaks, *Environ. Res. Lett.*, 2008, **3**, 024004.
46. H. A. Romijn, *Energy Policy*, 2011; doi:10.1016/jenpol.2010.07.041.
47. O. L. C. Maioli, B. A. Knoppers and S. A. Azevedo, *J. Atmos. Chem.*, 2009, **64**, 159.
48. C. Oppenheimer, V. I. Tsanev, A. G. Allen, A. J. S. McGonigle, A. A. Cardoso, A. Wiatr, W. Paterlini and C. de Mello Dias, *Environ. Sci. Technol.*, 2004, **38**, 4557.
49. F. S. Silva, J. Cristale, P. A. André, P. N. N. Silva and M. R. R. Marchi, *Atmos. Environ.*, 2010, **44**, 5133.
50. S. J. de Andrade, J. Cristale, F. S. Silva, G. J. Zocolo and M. R. R. Marchi, *Atmos. Environ.*, 2010, **44**, 2913.
51. J. E. D. Cancado, P .H. N. Saldiva, L. A. A. Pereira, L. B. L. S. Lara, P. Artaxo, L. A. Martinelli, M. A. Arbex, A. Zanobetti and A. L. F. Braga, *Environ. Health Perspect.*, 2006, **114**, 725.
52. S. D. Bonner and C. J. Kucharik, *Proc. Natl. Acad. Sci. U. S. A.*, 2008, **105**, 4513.
53. J. Hill, E. Nelson, D. Tilman, S. Polaski and D. Tiffany, *Proc. Natl. Acad. Sci. U. S. A.*, 2006, **103**, 11206.

54. M. A. Thomas, B. A. Engel and I. Chauby, *J. Environ. Eng.*, 2009, **135**, 1123.
55. B. Azhar, D. B. Lindenmayer, J. Wood, J. Fisher, A. Manning, C. McElhinny and M. Zakaria, *Forest Ecol. Manage.*, 2011; doi:10.1016/j.foreco.2011.08.026.
56. T. D. Mechan, A. H. Hulbert and C. Gratton, *Proc. Natl. Acad. Sci. U. S. A.*, 2010, **107**, 18553.
57. J. Fargione, *Proc. Natl. Acad. Sci. U. S. A.*, 2010, **107**, 18745.
58. M. J. Groom, E. M. Gray and P. A. Townsend, *Conserv. Biol.*, 2008, **22**, 602.
59. C. Wrobel, B. E. Coulman and D. L. Smith, *Acta Agric. Scand., Sect. B. – Soil Plant Sci.*, 2009, **59**, 1.
60. D. A. Landis and B. P. Werling, *Insect Sci.*, 2010, **17**, 220.
61. F. M. V. Cavalho, P. de Marco Jr. and L. C. Ferreira, *Biol. Conserv.*, 2009, **142**, 1302.
62. F. Hellmann and P. H. Verburg, *J. Environ. Manage.*, 2010, **91**, 1389.
63. J. Eggers, K. Tröltsch, A. Falluci, L. Maiorano, P. H. Verburg, E. Framstad, G. Louette, D. Maes, S. Nagy, W. Ozinga and B. Delbaere, *GCB Bioenergy*, 2009, **1**, 18.
64. S. Pfister, A. Koehler and S. Hellweg, *Environ. Sci. Technol.*, 2009, **43**, 4098.
65. S. Pfister, P. Bayer, A. Koehler and S. Hellweg, *Environ. Sci. Technol.*, 2011, **45**, 5761.
66. Y. Chiu, S. Suh, S. Pfister, S. Hellweg and A. Koehler, *Int. J. Life Cycle Assessment*, 2011; doi:10.1007/s11367-011-0328-0.
67. R. Murphy, J. Woods, M. Black and M. McManus, *Food Policy*, 2011, **36**, 552.
68. V. Natanelov, M. J. Alam, A. M. McKenzie and G. van Huylenbroek, *Energy Policy*, 2011; doi:10.1016/j.enpol.2011.06.016.
69. J. C. Clancy, *Eur. J. Dev. Res.*, 2008, **20**, 416.
70. M. Lehtonen, *Biomass Bioenergy*, 2011, **35**, 2425.
71. J. F. McCarthy and R. A. Cramb, *Geogr. J.*, 2009, **175**, 112.
72. J. S. H. Lee, L. Rist, K. Obidzinski, J. Ghazoul and L. P. Koh, *Biol. Conserv.* 2011; doi:10.1016/j.biocon.2011.07.006.
73. Y. H. P. Zhang, *Process Biochem.*, 2011; doi:10.1016/j.procbio.2011.08.005.
74. N. Sakar, S. K. Gopsh, S. Banerjee and K. Aikat, *Renewable Energy*, 2012, **37**, 19.
75. L. Reijnders, *Resources Conserv. Recycl.*, 2008, **52**, 653.
76. N. Kauffman, D. Hayes and R. Brown, *Fuel*, 2011; doi:10.1016/j.fuel.2011.06.31.
77. G. Centi, P. Lanzafame and S. Perathoner, *Catalysis Today*, 2011, **167**, 14.
78. T. Damartzis and A. Zabaniotou, *Renewable Sustainable Energy Rev.*, 2011, **15**, 366.
79. J. J. Cheng and G. R. Timilsina, *Renewable Energy*, 2011; doi:10.1016/jrenene.2011.04.031.
80. F. You, L. Tao, D. J. Graziano and S. W. Snyder, *AIChE J.*, 2011; doi:10.1002/aic.

81. E. Gnansounou and A. Dauriat, *Bioresource Technol.*, 2010, **101**, 4980.
82. R. M. Swanson, A. Platon, J. A. Satrio and R. C. Brown, *Fuel*, 2010, **89**, 511.
83. D. Bello, *Sci. Am.*, 2011, **305**(2), 58.
84. M. A. Carriquiri, X. Du and G. R. Timilsina, *Energy Policy*, 2011, **39**, 4222.
85. S. Spatari, D. M. Bagley and H. L. MacLean, *Bioresource Technol.*, 2010, **101**, 654.
86. A. Singh, D. Pant, N. E. Korres, A. Nizami, S. Prasad and J. D. Murphy, *Bioresource Technol.*, 2010, **101**, 5003.
87. S. Kim and B. E. Dale, *Biomass Bioenergy*, 2005, **29**, 426.
88. F. Cherubini and S. Ulgati, *Appl. Energy*, 2010, **87**, 47.
89. S. Fahd, G. Fiorentino, S. Melklino and S. Ulgati, *Energy*, 2011, doi: 10.1016/j.emergy.2011.08.023.
90. C. S. Goh and K. T. Lee, *Renewable Sustainable Energy Rev.*, 2011, **15**, 2714.
91. H. Blanco-Canqui and R. Lal, *Crit. Rev. Plant Sci.*, 2009, **28**, 139.
92. R. Lal, *Soil Tillage Res.*, 2009, **102**, 233.
93. X. Dong, M. Dong, Y. Lu, A. Turley, T. Jim and C. Wu, *Ind. Crops Prod.*, 2011, **43**, 1629.
94. S. Bhattacharjee, S. Datta and C. Bhattacharjee, *J. Cleaner Prod.*, 2006, **14**, 497.
95. W. Hanjie, F. Penning de Vries and J. Yongcan, *J. Environ. Sci.*, 2009, **21**, 488.
96. C. Xiao, R. Bolton and W. L. Pan, *Bioresource Technol.*, 2007, **98**, 1482.
97. R. Antikainen, R. Haapanen and S. Rekolainen, *J. Cleaner Prod.*, 2004, **12**, 919.
98. W. Jawjit, C. Kroeze, W. Soontarum and L. Hordijk, *J. Cleaner Prod.*, 2007, **15**, 1827.
99. G. Eggleston, Future sustainability of the sugar and sugar-ethanol industries, in *Sustainability in the Sugar and Sugar-Ethanol Industries*, ACS Symposium Series, American Chemical Society, Washington DC, 2010, 1–19.
100. L. Reijnders and M. A. J. Huijbregts, *J. Cleaner Prod.*, 2008, **16**, 477.
101. Q. Zhang, Z. Yang and W. Wu, *J. Sustainable Agric.*, 2008, **32**, 137.

Subject Index